T0149644

FOOTPRINTS

A PRACTICAL APPROACH TO ACTIVE ENVIRONMENTALISM

JIM BURHO

FOOTPRINTS
A PRACTICAL APPROACH TO ACTIVE ENVIRONMENTALISM

iUniverse books may be ordered through booksellers or by contacting:

iUniverse
1663 Liberty Drive
Bloomington, IN 47403
www.iuniverse.com
1-800-Authors (1-800-288-4677)

ISBN: 978-1-5320-7892-7 (sc)
ISBN: 978-1-5320-7893-4 (e)

Library of Congress Control Number: 2019911386

Print information available on the last page.

iUniverse rev. date: 08/13/2019

DEDICATION

In loving memory of my parents Vernis and Farrell
whose example of environmental awareness
laid the foundation for this book.

To my loving wife Sheila
who greatly supported and assisted me through
this effort of many long hours, days, and years.

To my many family and friends who took an interest in this
project and helped me with the extensive research required.

And to my daughter Jamey
whose invaluable encouragement and
editing expertise made this happen.

I also dedicate this book to all those who seek to
understand the environmental issues of our time with an
open mind and debate them through civil discourse.

CONTENTS

PREFACE

Smokey Meant Well But Beware of Talking Bears, And Other Early Environmental Experiences

"Seldom, very seldom, does complete truth belong to any human disclosure; seldom can it happen that something is not a little disguised or a little mistaken."
—Jane Austen, Emma

The earliest memories of my childhood are of being outdoors. Both my parents were active their whole lives in the scouting programs. Family vacations and weekend outings were almost always spent in tents, canoeing in the Boundary Waters of northern Minnesota, shivering while swimming in many of Minnesota's 10,000+ lakes in July and August, trying to catch the big one in those lakes, and hunting Ruffed Grouse in some of the most beautiful and aromatic forests a fall day can provide. Time around the campfire with family and friends dominate those early memories. I was fortunate to have been raised in the "land of sky blue waters," by parents who truly enjoyed nature and who continually reminded us that we humans had to take care of it. I can remember my father's wish for me was that the grouse remained plentiful and that I could enjoy the hunt with my children as he has. His wish has been fulfilled. Now I have that same wish for my children and their children.

As a young Cub Scout, I had my first encounter with Smokey the Bear, who taught me the precautions required of campfires and making the good point to "Drown your campfire out, dead out!" He became my first real symbol of environmental conservation. He was revered by all us young scouts. "Only you can prevent forest fires!" I don't know who put those words in Smokey's mouth; I suspect it was some well-intentioned U.S. Forest Service employee, and of course the point of those words was valid and not lost on me. "Don't let your carelessness with fire start a forest fire." But the message to this young Cub Scout from Smokey was that forest fires were bad, very bad. They killed animals, destroyed trees and plants, caused soil erosion, and fouled streams, and we humans were to blame for them.

It was many years later that I learned Smokey was not telling me the truth, or at least the whole story. Many forest fires, in fact, are not started by humans; but by nature, specifically lightning, earthquakes, volcanoes, and drought; and I really could do little to prevent those fires. I learned that at any given time there are many natural wildfires occurring continuously on the planet. I was further surprised when I learned that forest fires are, in many cases, actually good for the environment. Though fire does have its negative impacts, it also has positive effects which are very much part of the natural ecological cycle. It takes about ten years for a downed tree to decay and recycle on its own. Forest fires are essential in speeding up this cycle of life and decay as it allows the **biosphere** to turn over much more quickly and efficiently. After a forest fire there is a period of regrowth, very important when vegetation reappears. This period provides food for deer, moose, and elk, among many more animals which eat low lying shrubs, grasses and flowers; plants which disappear as the forest grows and prevents sunlight from reaching the ground. Without fire, you would have fewer of these early "transitional" plants, and far fewer of these animals. Some plants even require the heat of fires to germinate their seeds and then need strong sunlight to grow. Trees, such as lodge pole and jack pine, also rely to a certain extent on the heat of fire to open their cones and permit germination, In the aftermath of major fires, what remains in the form of stumps and vertical tree trunks play a role as shelters and a source of food for the wildlife habitat. The study of the impact of fires on the environment is a good way to understand how Mother Earth in the micro sense adapts to those forces attempting to throw off the "balance" of nature.

Why didn't Smokey the Bear put forest fires into perspective for this young Cub Scout? I should have been told all this. For sure he taught me well to extinguish my campfires for all the right reasons, but an equally good lesson I have taken from this is to challenge the words and "directives" of those in positions of authority, who for all intents and purposes, would have me behave in a manner of their choosing. The environmental community is full of Smokey the Bears on "both sides of the fire." Forewarned is informed.

The Wisdom of my Mother and Father

My mother and father were true and caring active environmentalists. I was raised surrounded by a family that was concerned about taking care of our environment. I was taught to drown out my campfires and leave all campsites in better condition than I found them. I first heard "leave no trace" at an early age from my parents. We never left a campsite without lastly "policing" it; littering anywhere was not allowed. They set the example of stewardship of the environment in conserving fuel by driving more slowly, keeping trash out of the lakes and rivers, and generally picking up after themselves. My father was frugal and conserved the few assets that he owned and those he shared with others. I was admonished more than once that "nothing is free" if I wasted or abused something "given" to me. But he also taught me to wisely use natural resources available to me to make my life easier, more productive, and safer. He was a master at improvising by using the resources around him wisely.

One of the great advantages of growing up in a small town, surrounded by state and federal lands, is that my brother and I and our friends had real access to the out-of-doors. Few "No Trespassing" signs had to be dealt with. Other than the normal cuts, scratches, or the rare broken bone, parental fears of allowing their children unsupervised access to the great outdoors were minimal or at least not an overt concern made known to us. For the most part we were encouraged to use our spare time camping, exploring, and hunting and we rode our bicycles to access those activities. Rarely were we ever "dropped off" from an automobile. This is not to say safety was put on a lower priority, but it certainly was made our responsibility.

We would ride our bicycles to go hunting and fishing generally with friends as the buddy system was the rule. As young "hunter-gatherers" the prime targets in those early years with those more primitive weapons were discarded tin cans, bottles, and later squirrels and rabbits. I guess I began hunting unsupervised around the age of twelve, though hunting with any of my friends was never as rewarding and satisfying as those hunts with just my dad and brother. While hunting with dad, no "creatures of opportunity" were ever bagged. Dad knew of our juvenile acts, innocent as they may have been, but he had a no tolerance attitude toward the reckless taking the life of our helpless furry and feathered friends of the forest—if you killed it, you ate it and that was the way it was when we hunted with dad.

In the lean years of the Ruffed Grouse population cycles, I almost got a sense from dad, after seeing him shoot one (most always in flight) that he felt badly after picking up his fallen friend. He often referred to his favorite game bird with great pride and respect by its official name "Bonasa Umbellus." The Ruffed Grouse was more to dad than "food on the table." To him it was his good reason to get out into nature and just enjoy it. When I hunted with him in his senior years I can remember him getting out of the car at a trailhead, getting his gun out of the trunk and before loading it taking a couple of deep breaths of that crisp fall forest air then stating, "I'm okay now!" As an adult, that certainly became my creed as well. Bagging the game was just an extra to the experience of "the hunt."

At the age of eleven or twelve on an overnight camping trip with my scout troop, I received my first scolding and first lasting lesson on conservation from someone other than my mom or dad. I had my first new all steel, official Boy Scout hatchet (which I still own to this day) and of course I had to try it out. Instead of a fallen tree, I chose to fell a live poplar, just a four-inch diameter, insignificant young tree. They were everywhere in abundance and the prime tree the pulp cutters sought for harvesting and sending to the paper mills. My new hatchet performed wonderfully, as I felled that young sapling with minimum whacks. When I yelled the customary "timber," it caught the attention of my scoutmaster who immediately appeared to investigate. I do not remember the exact words he used to reprimand me but the impact of whatever those words were has stayed with me ever since. I remember the point of his stern reprimand was I had just needlessly cut down a living tree; I had just wasted that tree. It would be the only tree I would ever cut down, just to cut it down.

Many years later I would hear a conservative radio talk show host proclaim, "No tree reaches its full potential until it is cut down." In spite of knowing the uses that timber would provide, when I heard this my thoughts immediately went back to my scoutmaster and wondered what he would have said to that comment.

The lesson I learned from needlessly cutting down that tree is the cornerstone of my search for dealing sensibly with the environment. How would human civilization have proceeded if it did not use the resources of the surrounding environment? It is human nature to want to be more secure, safer, and comfortable. Where else but from the environment can these needs be satisfied. How important to the "balance" of our environment is that one tree, that one gallon of gasoline, that one road, or that 100 kilowatts of electricity? Look what we humans have done with those resources! We sure have built some fine nests. It must be pointed out, however, that where access to those resources is limited or denied, humans live in some not so pleasant nests.

To understand an issue, my dad used to tell me, take it to the ridiculous. Okay, let's not cut down any trees! Let's not scar the Earth with mines, drill holes into it, or dam up its rivers, or exploit any resource to our advantage. Let's not light any fires to stay warm, for fear of emitting any CO_2 into the atmosphere or build great cities to scar the land and dense-pack the population. Let's promote a government that protects the environment with laws that forbid, forbid, and then forbid even more. Or, on the other hand, let's cut down every tree, strip mine the whole planet, drill holes everyplace on the planet where oil can be found, and let's use fire and combustion for any opportunity that will increase our immediate comforts. Let's pave every surface that suits our needs, dam every river to exploit its energy, and build roads that allow us access to anywhere. Let's promote a government that allows, allows, and then allows some more.

By pondering such thoughts at these extremes, it is clear that our lives at these limits would be significantly different, most likely, unsustainable. Do you get my point? Would anyone want to live at either one of those extremes? Of course not. Are we humans in agreement on where within the continuum of those extremes we want to live our lives? Not hardly. The goal of this book is to find a common sense middle ground on which to make policy decisions. It is immediately obvious this middle ground will not be acceptable to all, but will be acceptable to a majority. Compromise will be the key to survival of mankind as well as sustaining the environment. Can any of us get 100 percent of our way with the environment? Of course not, but could we accept getting only 50 percent of our way? Let's explore further with an open mind.

INTRODUCTION

"When the search for truth is confused with political advocacy, the pursuit of knowledge is reduced to the quest for power."
—Alston Chase

In his book, "An Inconvenient Truth," former Vice President Al Gore presented a graph showing the similar shaped curves of temperature increases with rising CO_2 concentrations over the past 1000 years.[1] That graph showed about a 0.8 degrees C temperature rise in the last 150 years and became iconic as it set off the environmental alarms that the Earth was in trouble. In this book, I will address these alarms and give the reader a logical and scientific perspective on them.

The purpose of this book is to take politics out of the debate. I will attempt to objectively present the different sides of this worldwide discussion while holding each viewpoint accountable so the reader can have a broader view of the issue than what he or she just hears from colleagues, and the news media. In so doing, I have made an honest attempt to make this book "apolitical." With some subjects dealing with government agencies as in the chapter on the Environmental Protection Agency it becomes a difficult task, however, I made every effort in those instances to remain objective and factual in my presentation.

This book has been more than five years in writing. My goal was to focus my research and writing in an attempt to make sense of the polarizing and politicizing of the environmental issues of conserving our natural resources while being good stewards of our environment. It grew into a vast panoramic of seemingly endless issues that I felt needed to be addressed. I have spent those last five years in research just trying to "get my arms around" the issues and make some sense of the often conflicting streams of "facts" with which we are flooded amid each new study, conference, and report on the complicated issues affecting our environment. My goal was to present the subject issues free from political interference. The more I researched, the more I became frustrated and thus convinced that this task could never be accomplished. The "experts," primarily those writing the literature, seemed to be agenda-driven, refusing to look at other sides of the environmental discussion, or even have that discussion, further complicating an increasing political penchant for intolerance of opposing points of view. Along with this our national political dialogue has become about as polarized as it has ever been, probably since the Civil War. This polarization of national politics goes along with and affects the concerns of **global warming** and **climate change**.

The stakes are high with these issues. For sure we as humans have become comfortable with automation, high tech gadgets, living in high rises, traveling with the convenience of high speed air, rail and interstate highways in vehicles so advanced all of which thirty years ago we could not have imagined. Through the development of the jet engine, most of Europe is within five hours of New York and no more than fourteen hours from any other place on the planet. How much of this "quality" of life would we give up to protect the environment from "disastrous" future consequences? What costs would we bear to this end? And what if, on the other end of the spectrum, even our most sacrificial measures could not effect a significant change in the direction our environment is progressing?

A major issue for all of us is that we have a predisposed tendency to "believe what we want to believe" and shut out trying to reason with opposing views. If we dismiss a point made by a disagreeable source, we are labelled "being political." To this point I realized a way to take politics out of the discussion. *Argue the point, not the source.* When we hear someone argue a point by saying "...well this guy is funded by big oil," or someone else argues a point claiming it "is funded by government/university grants," the *point*, then, is not being argued, rather politics is being introduced. If you don't agree with what Al Gore says, argue what he says, don't dismiss it just because Al Gore says it. If you don't agree with what Rush Limbaugh says, argue the point he is trying to make; don't dismiss it just because Rush says it. To dismiss or denigrate a point because it was made in the *Guardian* or on *Fox News* is a political rejection. It is neither logical nor intellectual, and certainly not scientific. A position that argues a point is likely the more credible one. Such discourse elevates both the intellectual level and the civility of the discussion and pretty much takes politics out of the argument.

This book will be that attempt to remove the protecting cloaks we all have and shed light on the opposing views through a broad and politically neutral perspective of the vast issues confronting us on the environment. If you are one who is convinced that the environment is on a path doomed by human influence, on a path not affected by human influence, or just not sure, this book is for you.

For some, the environment may be as significant as the air that's breathed while headed out of one's apartment on a crowded city street anywhere in the world; for others it may be the joy of dipping a canoe paddle in the pristine waters of the Boundary Waters Canoe Area of northern Minnesota, a blue water sail out of sight of land, or for others taking an off-road trip across the vast public lands of the western USA or Australia. For sure it must be the pleasure that comes from keeping one's self and family warm in winter and cool in summer with the flick of a switch or the striking of a match.

By any reflection, this is a huge topic that affects every aspect of modern humankind, wherever they live. My research in this book took me deep into the science of climate, an enormous scientific field requiring an understanding of astronomy, solar physics, geology, **geochronology**, geochemistry, **sedimetology**, **tectonics**, **paleontology**, **paleoecology glaciology** climatology, **meteorology**, oceanography, ecology, archaeology, and history and its related fields of physics, biological processes, thermodynamics, and oceanography. For sure I claim no expertise in any of these fields, but my research has opened my eyes to an understanding of the complexity of this topic as well as the climate's chaotic nature.

In my research I found most "climate scientists" admit the science of climate is not well understood.

This book is not for scientists, but I do expect their analysis. This book is for the "common" citizen of the planet, written with their everyday concerns and best interests in mind from a standpoint of how they react in their daily lives within the environment they both depend upon and the responsibilities they bear in safeguarding it. Somewhere in this discussion and debate, "common sense" perspectives should be considered. In the end, the scientist and the politician will have to deal with the concerns of the "common" citizen when they need the money to pursue their goals both in research and policy implementation. This book should open the door to awareness in the common citizen who will ultimately bear the costs and consequences in the direction the scientists and politicians would lead us. This is no time for the common citizen to be ignorant of those consequences.

I have always been an environmentalist. In my thirty-seven years in professional aviation I have seen more than most of our planet. I have had a unique opportunity to see how vast and beautiful it truly is and have, over the years, become incredibly mindful of its environment. My interest, however, to write this book was piqued while visiting relatives in Kansas on one of our annual father/son/daughter pheasant hunts one fall. On Uncle Logan's desk I noticed a magazine cover that jumped out at me. On that cover was a picture of a rustic, hard-working Kansas farmer with his boot propped up on his fence line, somebody's "feed and seed" ball cap on his head, holding a wheat stem in the corner of his mouth. The caption: "I am an active environmentalist, not an environmental activist. I became energized to pursue this angle. This book makes me an **active environmentalist** as well, with a keen interest in practicing both conservation and preservation of the environment I have enjoyed my whole life and want to pass on to my family and yours for generations to come. I will, for the most part, leave **environmental activism** to those interested in political and/or social movement agendas.

At the time of the Earth's creation, certainly nothing "lived" on this planet. Whether one believes in our religious teachings on creation or in the evolutionist's teachings is beyond the scope of this book. At first, we believe, the Earth was very hot and volcanic. Scientists can only make educated guesses that a solid crust formed as the planet cooled. As the planet continued to cool, water filled the basins that had formed in the surface, creating oceans. Through earthquakes, volcanic eruptions, and other factors, the Earth's surface slowly evolved into the shape that we know today. Its mass provides the gravity that holds everything together and its surface provides a place for us to live. But the whole process would not have started without the sun, our ultimate energy provider.

The Earth is fundamentally a different place now than it was at its creation so many years ago. Throughout its history, our Earth has been in constant change, constantly going through environmental changes, some of which encouraged and supported some forms of life, while at the same time caused the extinction of other forms of life. Ice Ages have appeared only to melt away as global warming ended each advance, forcing glacier withdrawals and a scarring of the earth left in the paths of their retreat. Hurricanes, typhoons, tornadoes, tsunamis, floods, dust and sand storms, forest fires, volcanoes, earthquakes, and massive soil erosion were common environmental occurrences, all before the "hand of man" ever set foot on this planet. In the macro sense, we would like to envision the earth in a sort of balance or equilibrium. In the micro sense, this Earth's environment has never been, nor should it ever be "in balance." It is constantly dealing with the laws of physics requiring reaction to forces from both within and from without our planet. I suggest the environmental scale of this planet was, is, and always will be out of balance, if only slightly, always being subjected to forces and adjusting to them. The human footprint has also left its mark on the planet both environmentally and physically. One only has to look at the small hamlets and major cities, the vast amounts of concrete and asphalt humans have laid down to connect them, the rivers they have dammed, and the vegetation altered and nurtured all in an attempt to improve our "quality of life."

This is a book that attempts to put into perspective the **forcings** that affect the Earth's environmental changes and seeming sense of balance it maintains as a result of those forcings. It is easy to point blame at humans themselves as one of those forcings affecting environmental changes, and certainly we humans have been one of the forces the environment must deal with. Our planet is about 78 percent water. The remaining 22 percent constitutes varying degrees of human ability to inhabit. On this immense planet that covers about 197 million square miles, and bombarded daily by an endless supply of solar energy from our own sun and cosmic energy from the universe as well, we must ask ourselves how much impact we humans have who only inhabit such a very small portion of it? This book will attempt to try to answer that question, or at least give it some perspective.

As I have said, I am not a scientist, but a concerned environmentalist. I am not driven by the agenda of industrialists or corporate capitalistic concerns that have much to gain in using the environment to enhance their bottom line. I am also not driven by political agenda and/or those who have much to gain from the political perspective. I am driven by the science and the search for the truth, and will use the science to make my points. I am also not trying to deprive the rich of their resource-gulping toys, and lifestyles. Nor am I trying to give back the forests to unregulated loggers. I have no political agenda. With few exceptions, I will take any notion of politics out of this discussion and concentrate on the science and logic in an attempt to better understand our environment and the forcings that disrupt any perceived balance. I want only the truth, but I'm also aware that the truth is not out there to be found in some well-defined form waiting to be discovered. Rather it is an elusive concept that only has meaning through broad understanding. I will draw heavily on the studies of as much of the continuum of scientific evidence on environmental issues that I can squeeze into this book and the thoughts of and writings of the scientists on all sides of this issue. I will compare and contrast the evidence of these scientists and though I will conclude with my own observations on that evidence, hopefully the way will be made easier for the reader to make her/his own better informed conclusions as well.

Sometimes this evidence may take the reader someplace that is uncomfortable or to views the reader would like to ignore (like that kid we all knew holding his fingers in his/her ears saying "lalala"). But we must remind ourselves that we cannot make the evidence what we want it to be or hope it should be. I will start with the basics and build on them and proceed to a conclusion—not the other way around. In my lifetime I have witnessed a replacement of the idea of truth, objective truth, with ideology. Ideology suggests political agenda and except for certain places where required for a degree of understanding, I will avoid ideology. It just simply "dirties" the discussion. I will dedicate one chapter, chapter 11, to the politics of all this in an attempt to show how political viewpoint affects environmental issues and policy.

So why would a non-scientist be interested in writing this book? The reason is simple: "Fixing" or "neglecting" our environment will impact our quality of life through its associated costs, no matter how one frames the argument. The scientists will not pay for their research or their findings or their recommended solutions. We the ordinary guy on the street will have to. Just as someone in our early life tried to warn us about those who would try to sell us something we didn't need, or didn't tell us enough of what we needed to know that was in our best interest; this book is that same warning. This book should allow that common person a fresh look at what is "fed" to them by a sales pitch from those with conflicting agendas and try to help them make more informed decisions.

This tome has been no easy task. The field is large, the research and literature vast, and the issues seemingly endless, but I will try to interject some common sense, as unbiased as I can make it to promote an unbiased and logical understanding of all the "environmental noise" "polluting" our daily lives. With no agenda other than a search for the truth, I wear no "blinders" in this research, and ask the reader as well to take off any blinders they may be wearing on these environmental issues.

CHAPTER ONE

HISTORICAL ENVIRONMENTALISM IN NORTH AMERICA

"Humankind has not woven the web of life. We are but one thread within it. Whatever we do to the web, we do to ourselves. All things are bound together. All things connect."
—Chief Seattle, Chief of the Duwamish and Suquamish tribes

The Native American Indian civilization which was well-established and as early as 1000 BC[1] is where I will begin the history of environmentalism in North America. As I presented in my Preface, the culture of those Native Americans has become to be revered, respected, and emulated as a culture that was one with nature and lived in harmony within its environment. "When the Indians killed an animal," I was told, "they used every bit of that poor slaughtered critter to make their lives easier." Unlike the early settlers who massacred the buffalo, sometimes shooting them out of the windows of trains in which they were riding, occasionally not even for their hides, but just for the "thrill" of the moment.

Then in college, when I studied more closely our Native Americans, I found out this culture, which I was lead to believe promoted all those wonderful things, actually wasn't so environmentally friendly in many respects. "The spiritual connection attributed to Native Americans frequently does not mesh with the history of Indian resource use," said Terry Anderson in an article published by the Property and Environment Research Center (PERC).[2] "To claim that the Native American Indian lived without disturbing the balance of nature is akin to saying that they lived without touching anything, that they were a people without history."[3] When the Native American needed hides to stay warm and cover the frames of their teepees and lodges, often they would drive or stampede herds of buffalo off cliffs to kill them, skin them en masse leaving the meat to rot on the plains leaving it all to the scavengers to make good use of the carcasses. I saw this first hand later in life while stationed with the United States Air Force in remote northern Alaska, along the Chukchi Sea. I saw our Native American cousins, primarily for commercial gain, slaughter caribou, polar bear, and whales, ignoring any limits placed on them by the government; claiming it was their right as a "sovereign" nation to do with the environment as their culture dictated, without regard to the wishes of the outside world.

I've seen the salmon population in the northwest decimated by our Native Americans, as they would stretch their nets across rivers trying to catch as many of the salmon as they could as this beautiful fish was going up stream with the sole purpose of making more little salmon when they got to their birthplaces in those streams and their tributaries. Such type of fishing is not allowed to us native Americans who only have five or six generations of "native" to claim.

When I lived in Alaska I got to know the villagers in some of the smaller coastal villages because I owned and flew my own small airplane while there. I found out that us non- "native" Americans could get a hunting license to kill a polar bear but only one license per every two generations. I could get a license to kill one, but my children could not. In the spring of my year stationed there, I personally saw in the cold tundra air racks upon racks of polar bear hides drying which would soon be crafted into the ritual needs of the Inuit culture and some into tourist souvenirs which brought them cash. I observed that these natives rarely spoke about conservation and preservation of the polar bear, whose only natural threat is man, believing the Great Spirit would always provide more. They felt the same about the beluga whale as they hunted, killed and hauled them up on the ice cap to "jerk" in the sun, in what appeared to disregard the "one per village" limit imposed on them.

The early Native Americans who farmed by extensively clearing and deforesting (burning) the land and then when soil fertility was diminished moved on to clear more land and start the process over again. Similarly when game was plentiful only the choicest edible cuts were used leaving the rest to rot. Samuel Hearne, a fur trader near Hudson's Bay, recorded in his journal in the 1770s that the Chipewyan Indians would slaughter large numbers of caribou and musk ox, eat only a few tongues, and leave the rest to rot.[4] These early native Americans also manipulated the land to improve hunting, burning forests and undergrowth to promote new vegetation to attract game animals. "Indeed, because of this burning, there may have been fewer "old growth" forests in the Pacific Northwest when the first Europeans arrived than there are today…." and "these human-caused fires altered the succession of forests." In the Southeast, for example, oak and hickory forests with a higher carrying capacity for deer were displaced by fire-resistant longleaf pine which supported only limited wildlife."[5]

One only has to drive through the Navajo Nation along the 100 mile stretch of U.S. 160 in northern Arizona. It is the most littered section of a U.S. highway I have seen. It might be easy to blame the U.S. government for "putting" Native Americans on reservations to begin with as the cause, however, I think this would be a bit of a stretch if we are going to give the Native American culture credit as caretakers of their environment.

I don't mean to portray the culture of Native Americans disrespectfully, just give it perspective. They were trying to survive and prosper in a harsh environment. They did what they had to do, and in spite of how we view sustainability today, left a respectable legacy and continuing culture.

CHAPTER TWO

THE BASIC SCIENCE

"The Atmosphere is among the earth's most complex dynamic systems: subtle in its chemistry, chaotic in its flow. It interacts with everything from the solar wind to the deep oceans. It is subject to insults great and small, brief and enduring, from men and meteorites, volcanoes and termites, wildfires and algae blooms-a list without end."
–Russell Seitz, Senior Research Fellow at the Climate Institute

Introduction

What is the environment? For the purposes of this book, I will define the environment as, "the surroundings or conditions in which a person, animal, or plant lives." The environment may be considered in the micro sense as a specific location on Earth or it may be considered in the macro sense with a more global perspective. Due to the limitations of scope, this book will not delve extensively into the environment of the ocean depths or lakes, but will include their impact on a broader perspective as they affect landmass environmental issues. In the macro sense for human concerns, we have a continuum from temperatures of heat and cold not suitable for the most part to plants or animals, to the lush tropics which teem with life of all sorts. This also includes the impact of altitude and atmospheric pressures from the lowest dry land on Earth, the shoreline of the Dead Sea at 1,354 (413m) below sea level, to our highest point on Earth; 29,035 ft (8850m), Mt Everest.

With all the conversation on the apparent warming of the planet and possible causes, my overall interest is to sort out these causes and put them in both a logical and economical perspective without getting wrapped up in the politics of them. In so doing I will make some observations on the extent to which these forces affect our planet's vast environment.

In the micro view, the "**butterfly effect**" could well be the consequence when the mere impression of one's footprint in the forest or the mutation of a beetle, for example, could impact change on the overall environment to some extent. In the macro perspective, a whole host of forces exist from solar and cosmic activity, to the movement of the Earth's crust, to the static electricity caused by the movement of air masses, to the **Industrial Revolution** which resulted in the massive use of fossil fuels which increased carbon emissions into the atmosphere, to mention just a few. All these forces or events impact our environment. The vastness of our Earth and its atmosphere is difficult to comprehend and our planet and its environment is but a small "bug spot on the windshield of the universe." I think this bears keeping in mind as we look at the forces that impact the environment in which we live.

Just as the continuum of affecting forces upon our environment is vast, so are the views of each of us who inhabit this Earth. It is important to remember how attitudes and opinions on the environment will vastly differ among people living in dense-packed population centers, coastlines, desert/ arid land, etc, as opposed to those living distant from those conditions. We have some, at one end of the spectrum of those attitudes and opinions, who believe humans are the major force causing changes in the Earth's environment and specifically its climate (called **anthropogenic** causes), and others at the other end who believe the natural processes of the Earth, by themselves, are the major force affecting our environment. In this chapter I will present a review of these causes, both natural and anthropogenic. This assessment will lead to an overview of how our planet "fits" into its solar system and to a lesser extent the cosmos. I will present a view of how our atmosphere's unique makeup protects, sustains, and maintains life as we know it on planet Earth. I will define **greenhouse gases (GHGs)** and describe how they work, and discuss briefly their individual potencies, and how each are generated both from "natural" sources, and anthropogenic sources. Lastly I will explain how the atmosphere plays a major role in both the carbon and water cycles which provide and sustain life on Earth.

Discussion

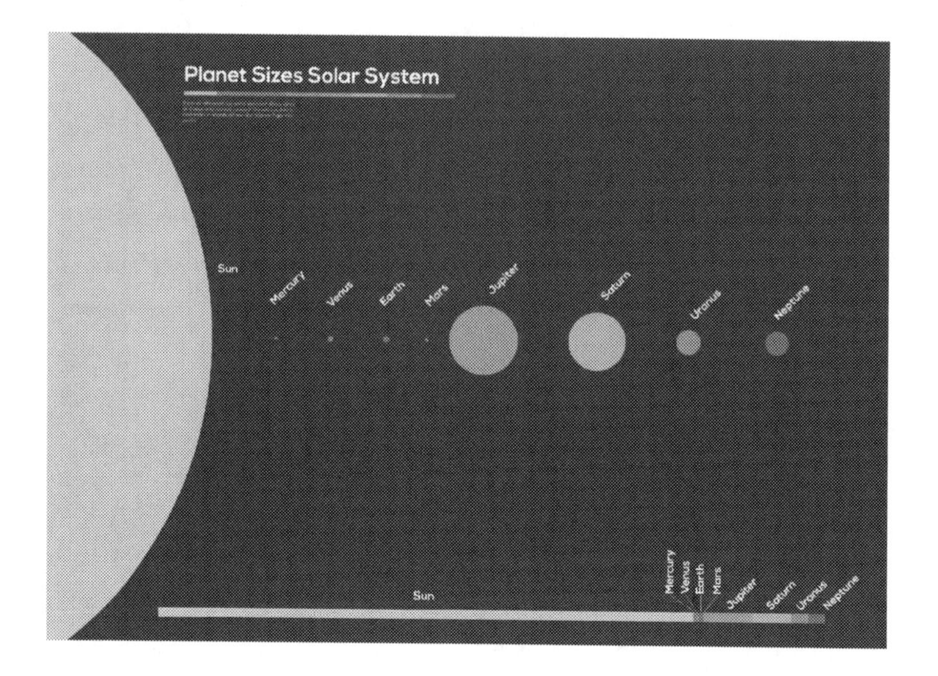

Here's some "gee whiz stuff," common knowledge to some of us, but not to most of us. It is a gross understatement to make the point that our sun has the major impact on our environment. Our Sun is enormous and as can be seen contains just about all of the mass in the solar system. At a distance of 93 million miles, it is 109 times wider than our Earth, and 333,000 times heavier. In contrast to our "crusted" Earth, the sun is a huge flaming ball of gas whose surface temperature is estimated to be about 10,000 degrees Fahrenheit and is for all practical purposes the sole source of heat for our planet. If not for our atmosphere, quite unique for all planets within our solar system, daytime temperatures would soar to unbearable temperatures and would drop to sub-freezing temperatures during nighttime. The moon, for example, which has no atmosphere to speak of and rotates in orientation to the sun once in every 27 days, has daytime temperatures of 225 degrees F and nighttime temperatures of -250 degrees F. If one could stand on the surface of the moon at sunrise or sunset, the temperature rise or drop would be dramatically felt and almost instantly.

By comparison, the surface temperature of Venus, at a distance of 67 million miles from the sun, is a whopping 840 degrees F; and Mars at 142 million miles from the sun has an average surface temperature of -63 degrees F. The fact that Earth's average surface temperature at 55 degrees F is comfortably between the relative extremes of Venus and Mars cannot be explained by simply saying our orbit around the sun is at just the distance to absorb just the right amount of radiation. It's also the result of having just the right kind of atmosphere which mitigates this rise and fall due to its insulating properties. The scientific reason for this is not completely understood primarily because climate models have not done well in predicting future global temperatures as we shall see in a later chapter. This is very important to understand when trying to deal with the causes of global warming and climate change.

The main focus of this discussion will center on the makeup of our atmosphere and the great influence it has on our environment.

Our atmosphere is currently made up of 78 percent nitrogen, 21 percent oxygen, .004 percent carbon dioxide, and less than 1 percent other[1] as depicted in the following graph:

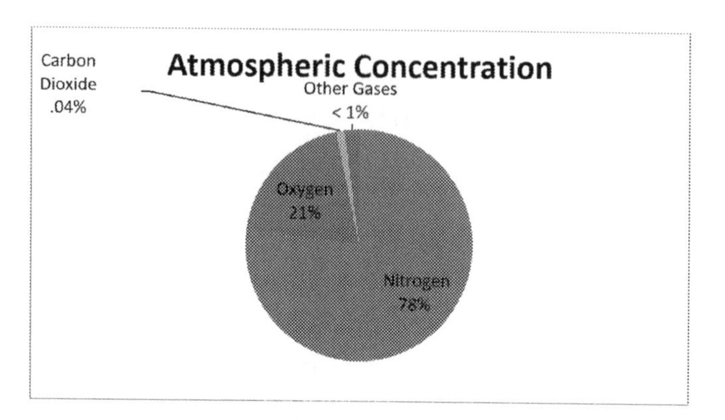

Historically it is important to note that the current composition of the atmosphere has not been constant over time:[2]

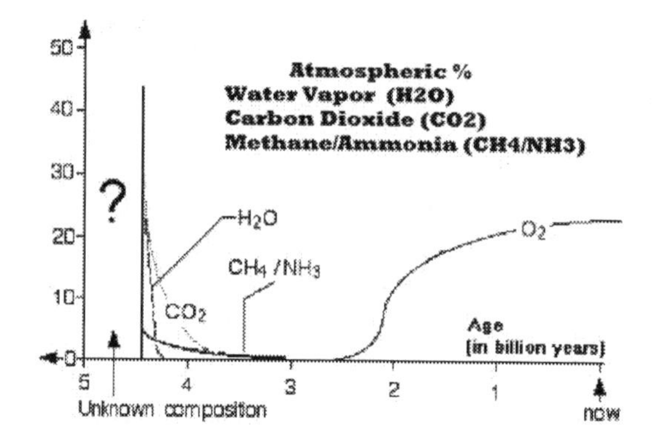

As will be discussed shortly, it is generally believed that neither oxygen nor nitrogen have any significant, or at least well understood, insulating or heat-trapping properties of their own. Within this "other 1 percent" are trace gases that absorb heat and are referred to as greenhouse gases (GHG). They are water vapor, carbon dioxide, methane, ozone, and nitrous oxide.[3]

These trace greenhouse gases create the greenhouse effect. Simply explained though complex in nature, short-wave radiation (solar high energy radiation) can pass through the clear atmosphere relatively unimpeded by the GHGs. Because the Earth is much colder than the sun, it radiates at much longer wavelengths (in the infra red spectrum). Not all of this long wave (low energy) terrestrial radiation emitted by the surface of the Earth escapes back into space but is absorbed (trapped) and re-emitted by the GHGs in the atmosphere. This absorption and re-emission which warms the lower atmosphere and the surface of the Earth is known as the greenhouse gas effect, without which the lower atmosphere would be significantly cooler. On average, the outgoing long wave radiation balances the incoming solar radiation, but both the atmosphere and the surface will be warmer than they would be without the greenhouse gases.[4]

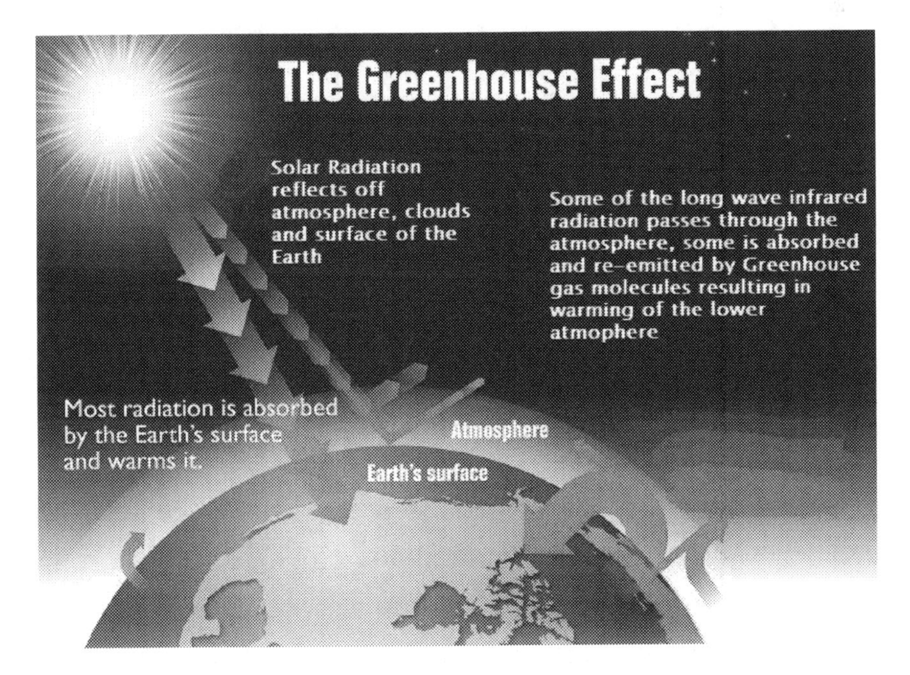

Just as the glass walls and roof of a greenhouse let the heat energy of the sun through, and then "trap" some of it from escaping, so do greenhouse gases allow the sun's energy to enter the atmosphere, and "trap" some of it from re-radiating back into space. It is important to note here that the sun's energy that penetrates the atmosphere and then subsequently is trapped is infrared energy. This entrapment occurs as these peculiar trace gases turn infrared radiation into vibration of GHG molecules as the radiation from the sun or reflection off the Earth's surface impacts them.[5] It is believed the unique character of these greenhouse gases exists because, unlike oxygen (O_2) or nitrogen (N_2), greenhouse gases have three or more atoms in their molecular makeup (water vapor—H_2Og, methane—CH_4, ozone—O_3, and nitrous oxide—N_2O) which allow for greater vibration when "bombarded" by infrared energy.[6] This heat entrapment, we call the greenhouse effect, refers to how humans contribute, through emission of these GHGs to the overall concentration of GHGs. The anthropogenic emission of GHGs into the atmosphere is the human component to any global warming.

As has been shown, 98-99 percent of our atmosphere is made up of mainly nitrogen, oxygen, argon and a few other trace gases. The remaining 1-2 percent are the GHGs. Of that small amount, water vapor makes up 95 percent of it and the remaining 5 percent are the trace GHGs of CO_2, methane, ozone, nitrous oxide, etc.

Depending on the source of study, 96-98 percent of greenhouse gas emissions annually occur naturally without human causation[7] through **biogenic** processes such as the decomposition of biological materials, forest fires, volcanic action, and fermentation, to mention a few. The remaining 2-4 percent of the annual CO_2 emissions is caused by human or anthropogenic sources which include: burning fossil fuels, deforestation, changes in wetland structure, along with other anthropogenic GHG emissions from fermentation of livestock waste, landfill methane emissions, fluorocarbon emissions, and agricultural activities using nitrogen-based synthetic fertilizers.

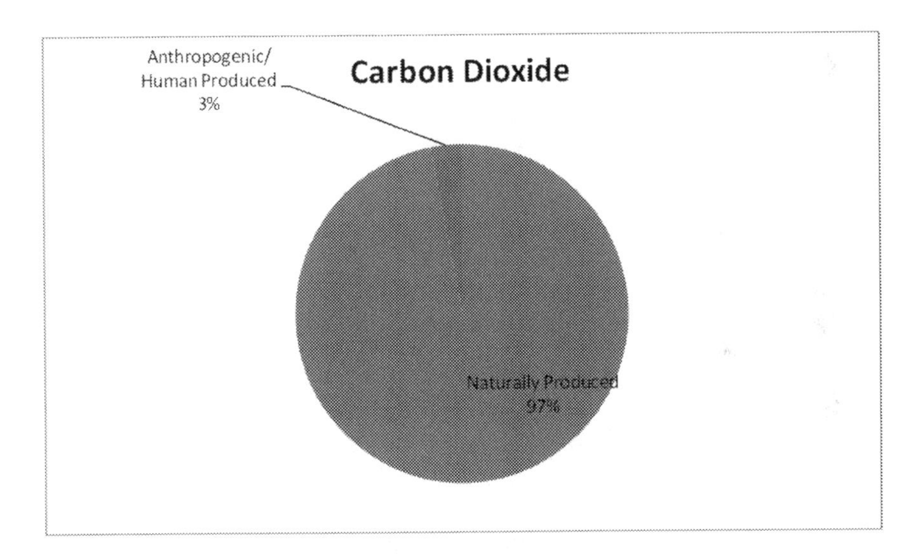

It is important to note that increases in CO_2, no matter the source, can lead to large accumulations over time because these CO_2 molecules can remain in the atmosphere for more than a century. The ratio, however, as shown in the graph above appears to remain relatively fixed.[8]

Water vapor, at 95 percent of GHGs in the atmosphere, is a constant. Of the remaining 5 percent only 3.6 percent of that 5 percent is CO_2. Putting the human contribution further into perspective, of the 3.6 percent total CO_2 concentration, only 3 percent is **anthropogenic** in origin.[9]

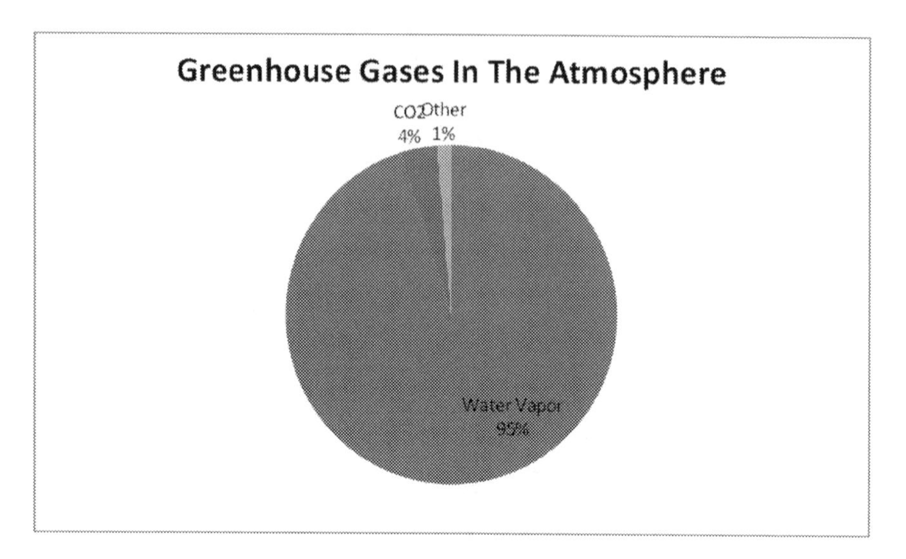

Let's review these GHGs with a brief review of their characteristics and impact on radiation entrapment:

Water Vapor (H_2O)

Water Vapor is the most abundant and constant GHG which comprises 95 percent of all greenhouse gases and is not produced by human activity. Globally this 95 percent is basically a constant amount varying somewhat with increases as temperatures warm and decreases as temperatures fall. When water vapor amounts increase as temperatures increase, this is called a positive feedback. A negative feedback would occur as temperatures decrease.

Carbon Dioxide (CO_2)

The Swedish chemist Svante Arrhenius was the first to quantify the atmospheric warming effect of CO_2 in emitting infrared radiation. In 1896 he published a paper showing how he calculated that a doubling of atmospheric CO_2 would cause a global temperature rise of 5 to 6 degrees C (Arrhenius 1896).[10] His calculations were rudimentary and not based on any understanding of the science of **spectroscopy**. Arrhenius's much quoted 1896 paper was subsequently incorporated in part into the UN's **IPCC** temperature projection models (see chapter 6). As this science was developed, the effect of the doubling of CO_2 on temperature increase went to 4.2 degrees C by climatologist James Hansen in 1988, and continuously adjusted downward. In the 1995 IPCC report it was decreased to 3.8 degrees C increase for a CO_2 doubling, and at the time of this writing adjusted down again to a 2.5 degrees C increase.

Carbon dioxide is emitted by both natural and human activities. It is produced primarily by the combustion of hydrocarbons (fossil fuels), the fermentation of organic material, and the respiration of humans and other animals. This greenhouse gas is the primary human contribution to global warming through the burning of fossil fuels that have powered the Industrial Revolution, electricity generation by power plants, and the transportation industry.

CO_2 is also emitted from volcanoes, hot springs, geysers, and other places where the earth's crust is thin. It is also found in lakes, at depths under the oceans, and also found mixed with oil and gas deposits. One source of CO_2 emissions—little recognized and almost unknown to most—other than to geologists, are the many unwanted coal fires which continue around the world and (unlike grass and forest fires) burn for years. These fires are the result of coal deposits igniting from lightening, grass or forest fires which continue to smolder underground after the surface fires have been extinguished. When coal burns underground, it does so incompletely therefore emitting another greenhouse gas, besides carbon dioxide, methane; as well as the toxic gas carbon monoxide. These coal fires are very difficult to extinguish as they propagate in a creeping fashion along mine shafts and cracks in geologic structures. Pictured below is the coal fire in Turkmenistan called "The Door to Hell," which has been burning since 1962.[11] It is a crater 325 feet wide and 65 feet deep.

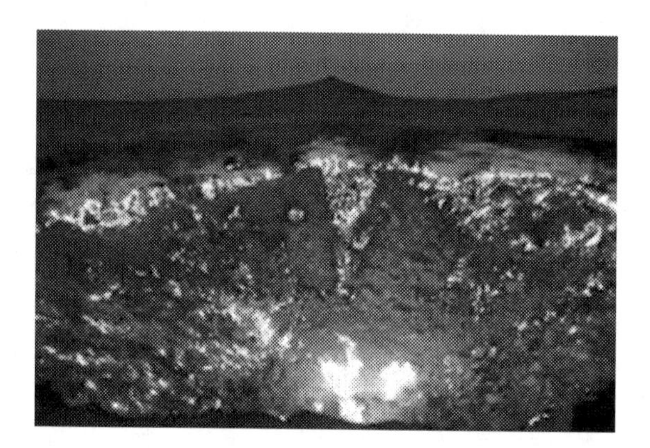

Carbon dioxide production from these unwanted coal fires around the world is enormous. According to one technical paper from the Department of Energy, about 2 percent of all annual industrial global emissions of carbon dioxide come as a byproduct of unwanted coal fires in China alone.[12] We have coal fires burning here in America, as well, including one near Laurel Run, Pennsylvania that's been burning since 1915.

I include these coal fires in the discussion on CO_2 emissions to add perspective on how varied the emission sources of this GHG are, which go largely unreported by the media or are difficult to find, even in science journals.

Methane (CH_4)

Methane is a very potent but rare greenhouse gas in our atmosphere, emitted by both natural and human activities. We can tell by its molecular structure consisting of five atoms that its potential for trapping heat would be higher than CO_2 with, a global warming potential 21 times greater than carbon dioxide according to the EPA.[13]

Methane is the main component of natural gas, and though rare in the atmosphere, it is probably the most abundant organic compound on earth. The relative abundance of methane along with its clean burning nature makes it an attractive fuel. Compared to other hydrocarbon fuels, burning methane produces less carbon dioxide for each unit of heat released and produces more heat per mass unit than other complex hydrocarbons.

Its emission sources, besides the burning of natural gas, are both natural and human related. Natural sources of this gas include wetlands, animal digestive processes, and the oceans. Human-related sources include landfills, livestock farming, as well as the production, transportation and use of fossil fuels. Among its many natural sources, wetlands are major methane emitters during the decay of organic matter as they provide a habitat conducive to methane-producing bacteria called **Methanosarcina** which produce methane during the decomposition of organic material in an environment lacking oxygen. Another natural source of methane emission, though aided by agriculture, is from animals as mentioned through their digestive processes (manure and flatulence).

Wild fires are also a source of atmospheric methane due to incomplete combustion of organic material during this process. Methane is also emitted from deforestation in tropical areas as it is released from the newly exposed soil, and in polar regions when permafrost melts and releases trapped methane within that soil.

Nitrous Oxide (N_2O)

N_2O (commonly known as "laughing gas") is a major GHG, not due to its quantity on earth, but rather on account of its high potency for trapping heat. Thus, despite its low concentration, N_2O is the fourth largest contributor to these GHGs and it can be emitted by both human and natural causes. It ranks behind water vapor, carbon dioxide, and methane.[14]

Agriculture is the main source of human-produced N_2O, through processes involved in soil cultivation, the use of nitrogen fertilizers, and animal waste-handling. The gas is emitted through natural causes by the natural nitrogen cycle as decaying vegetation produces various nitrogen compounds, N_2O being one of them.

Fluorinated Gases

Fluorinated gases are created by humans and do not occur normally in nature. They are used mainly in industrial processes. The creation and/or use of refrigerators, air conditioning systems, foams and aerosols are the main source of fluorinated gas emissions.

Again it should be noted: 98-99 percent of our atmosphere is made up of oxygen and nitrogen, leaving 1-2 percent for all the other gases including the GHGs. In the broad context of GHG concentrations, we are talking about some very small numbers.

How It All Works

The "...surface of the Earth acquires nearly all of its heat from the sun, and the only exit for this heat is through the door marked 'radiation.'"[15] The Earth travels around the sun in an elliptical orbit, elliptical because of the gravitational interaction with the other planets in our solar system. Our planet Earth spins on an axis tilted at an angle that varies between about 22.1 degrees and 24.5 degrees on a 41,000 year cycle and the Earth, on this tilted axis precesses (wobbles) on approximately a 20,000 year cycle.[16] These factors cause a variance in the distribution of the constant radiant solar energy received across the planet. It is this roughly 23 degree tilt of the axis of the Earth that causes the seasons and climate changes that occur as a result of the varying angle of the sun's rays. GHGs insulate the earth because of their ability to absorb and trap this radiation of solar energy. The varying makeup of the Earth's surface also has an impact on heat retention of our atmosphere. In the atmospheric greenhouse effect, the type of surface that sunlight first encounters is the most important factor. Its first encounter would be clouds and then the surface of the Earth to include forests, grasslands, ocean surfaces, ice caps, deserts, and cities all of which absorb, reflect, and radiate the sun's rays differently. When solar rays hit a dark surface, such as forests, heat is absorbed, resulting in increased temperatures. Solar rays reflecting off lighter surfaces such as clouds and the polar ice caps don't absorb as much radiation, but instead reflect more of it back into the atmosphere and space causing a relative cooling effect. As a result, it can be understood that the more energy reflected back into space, the greater will be the cooling effect. The term "**albedo**" refers to the reflective quality of a surface. The lighter the surface, the greater will be its albedo effect. The relevance of this term will be further discussed in a later chapter.

To understand why these GHGs act as effective global insulators, I will explain a few basic facts about solar radiation and the structure of atmospheric gases.

This graph of the electromagnetic spectrum[17] gives approximate percentages for solar radiation absorbed and reflected by the earth.

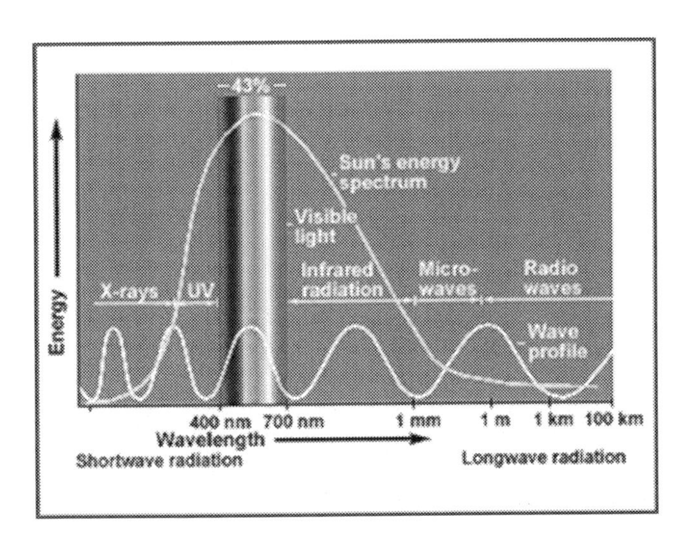

It might appear this term, along with the graph, is quite complicated but actually it is not. Very simply, electromagnetic energy is energy emitted by charged particles traveling at the speed of light (in a vacuum) and at different wavelengths depending on what generates this energy. Most of the energy radiated from the sun is in the visible or near visible portion of this spectrum, between 400 and 700 nm (nano meter—1 billionth of a meter in length), an incomprehensible length for most of us. The shorter the wavelength, the more energy it contains. You can see that on the backside of this visible spectrum are the ultra violet (UV) wave lengths, very energetic and capable of breaking apart molecules which in essence is what happens as this UV energy can cause sunburn and skin cancers.

As discussed earlier, these greenhouse gases are all molecules of more than two component atoms that are bound loosely enough to be able to vibrate with the absorption of infrared radiation. The major gases in the atmosphere are oxygen (O_2) and nitrogen (N_2) both two-atom molecules believed to be too tightly bound to vibrate and therefore not able to absorb heat.[18] For that reason they are believed not to contribute any greenhouse effect.

The characteristics of these gases were first observed or determined in the 1800s when the term "greenhouse effect" was first used in association with CO_2.[19] At that time it was used to explain the naturally occurring function of this trace gas, and did not have any negative connotation. It was not until the 1950s that the term began to take on an environmental concern for temperature change of our planet and began to take on a more negative and sinister connotation for its possible impact on enhancing the greenhouse effect and temperature. It is important to remember, in all the negatives of GHGs, that without the greenhouse effect, life on earth as we know it would not be possible.

In our discussion of the effects of greenhouse gases, I will focus mainly on CO_2, since it is the most prominent greenhouse gas we humans have some control over in terms of emissions. However, we must always allow for the consideration of those other trace GHGs, and remember that water vapor is the most abundant of them all.

The Carbon and Water Cycles

Carbon is fundamental to life. All living organisms are based on the carbon atom. This carbon atom has some unique characteristics in that it has the ability to form bonds with as many as four other atoms (including itself); and these compounds can exist as a solid, liquid, or gas under conditions commonly found on the Earth's surface. The attributes of the remarkable carbon atom make possible the existence of all organic compounds essential to life on earth.

Carbon atoms are constantly on the move through living organisms, the oceans, the atmosphere, and the crust of the Earth. This movement is called the carbon cycle. The paths taken by carbon atoms through this cycle are extremely complex, and it may take millions of years to complete the cycle.

Let's follow just one typical path taken by many of these carbon atoms starting, let's say, 350 million years ago. As part of that CO_2 molecule this carbon atom finds its way to the leaf of a fern plant in a tropical forest swamps during that period of time. Through photosynthesis, the oxygen from the CO_2 molecule is released back into the atmosphere and the carbon atom is used to build a molecule of sugar.[20] This sugar molecule is stored in the plant as part of its long-lived structure. Later when the fern dies and sinks into the muck floor of this swamp, it remains there for millions of years along with others and form a thick layer of dead plant material. Gradually as the climate changes (as it always does), and becomes warmer and less tropical, this layer is covered with sand and dust along with other materials sealing this ancient swamp with all its decaying vegetation beneath by an ever-thickening layer of sediment. Slowly this sediment layer turns to sedimentary rock. The carbon atom remains trapped in the long-vanished swamp while the pressure of the layers above slowly turns the material into coal.

Three-hundred-fifty million years later this sediment layer is discovered by humans and mined for its coal and burned to fuel the industrial revolution. The process of burning releases the energy stored in the carbon compounds in the coal and reunites the carbon atom with oxygen to form CO_2 again thus completing this part of the carbon cycle and starting a new one. Many other paths are possible, some taking only hours or days to trace, while others, like the one above, take many millions of years.

"The aggregation of the possible paths of carbon, where it may be stored for extended periods, where it is likely to be released to the atmosphere, and what triggers those sources, together defines the carbon cycle."[21] The places on our planet that store carbon are referred to as "sinks." The triggers that release this stored energy from carbon are called "release agents." These releasing agents can be man-induced through burning processes and agricultural uses, or through natural releases such as photosynthesis, volcanoes, and earthquakes just to mention a few.

For all the bad press CO_2 gets as the GHG which is warming the planet, it is a major nutrient to life on this planet, and without it life could not exist. This gas is required by plants in the process of photosynthesis which is the chemical action energized by sunlight to change water and carbon dioxide into the sugar the plant needs to survive and grow. The question before us environmentalists is: "can too much of a good thing be harmful"? I'm not suggesting that too much of a good thing is okay, but to label CO_2 a pollutant, as will be discussed in chapter 6, needs to be defined and addressed by the scientific community. For those interested in some more gee whiz, the photosynthesis equation is:[22]

$$6CO_2 + 6H_2O \longrightarrow C_6H_{12}O_6 + 6O_2$$

It can be seen here how important CO_2 is to this life-giving process. Add sunlight and we get sugar, oxygen and water, the products of photosynthesis:[23]

$$6CO_2 + 12H_2O + Light + C_6H_{12}O_6 + 6O_2 + 6H_2O$$

This magical process creates the air we breathe and the food we eat.

Obviously, plants are important, but not just because they give us food to eat and oxygen to breathe. They also regulate the amount of CO_2, in the atmosphere and protect the soil from wind and water erosion. Plants release water into the air during photosynthesis, and along with the rest of the water on the planet, take part in the huge water cycle, which is controlled by the sun.[24]

The water cycle much less complicated and not nearly so long in duration, is also critical to the environment and the heating and cooling of our planet. As we have previously mentioned, the most prominent and powerful of the greenhouse gases is water vapor. The water cycle, also called the hydrologic cycle**,** is a natural Earth process that exchanges moisture between bodies of water and land masses. The water cycle is responsible for clouds and rain which makes life possible on this Earth.

Briefly explained, here's what happens in the water cycle:[25] As the sun shines on the surface of bodies of water (small or large) it excites the water molecule, H_2O, causing it to evaporate and become water vapor, H_2Og, which then enters and rises into the atmosphere. Plants add to this water vapor through transpiration, the process through plants absorb water through their roots and then give off as a byproduct of photosynthesis water vapor through pores in their leaves, the whole process, of course, powered by the sun. In colder regions, water sublimates, or changes directly from ice into water vapor. The higher this water vapor rises, the cooler it gets, slowing down molecular motion, causing this water vapor to condense at the dew point as they cool and form clouds. The condensed water droplets combine inside the clouds, and when they get big and heavy enough, they fall as precipitation in the form of rain, snow, sleet or hail, depending on the temperature and other conditions (particles, e.g., pollution, in clouds can decrease the amount of rainfall by requiring droplets to be bigger and heavier before they can fall.). When this precipitation falls to the surface of the Earth the cycle is complete.

Briefly I will note here and explain more fully later, that the scientific understanding of cloud formation and the influence it has on the greenhouse effect is not well understood. Research suggests the Sun's cosmic rays play a large role in atmospheric cloud formation and thus surface temperatures, of which current modeling does not account for very well, if at all.[26]

Summary

As our Earth revolves about our sun, it is that sun which powers the processes that control our climate and the content of our atmosphere. Without it, we would not have oxygen or liquid water in our atmosphere, nor would we have varying weather or seasons. The impact we humans have on these processes and content, though highly probable, is not yet known and will be the focus of following chapters.

CHAPTER THREE

GLOBAL WARMING AND CO2, A HISTORICAL REVIEW

"Facts are stubborn things; and whatever may be our wishes, our inclinations, or the dictates of our passion, they cannot alter the state of facts and evidence."
—John Adams

"1998 was the hottest year in modern history."
—Vice President Al Gore and NASA Scientist James Hansen

Introduction

The subject of global warming is probably the "hottest" social and scientific topic of our time. In recent years as the apparent warming of Earth's atmosphere has slowed and not reached projected temperatures, the term "global warming" is used less in favor of the more general term, "climate change" (used with the understanding that the Earth is still warming and it is this warming which is driving the observations of climate change). I will use them interchangeably as they apply to the particular point I am making.

I will divide the perspectives on global warming into three camps. The first camp, **the climate warmists**, consist of those who believe humans are the driving factor contributing to global warming, primarily through the burning of fossil fuels which emit considerable amounts of CO_2 as well as other human-caused GHGs emissions into the atmosphere. They believe these anthropologic GHG emissions are the cause of observed global warming and climate change. They believe these emissions need to be greatly reduced before their concentration levels increase to a "tipping point" beyond which the planet's temperature will continue to rise without any natural ability to stop it.

The second camp, **the climate deniers**, consist of those who believe, increasing anthropogenic GHG emissions have little effect on global warming and that global temperature changes are primarily the result of natural forces at work on the planet. This term was originally introduced by the climate warmists to associate this group to holocaust deniers. Though a derogatory label in that regard, it has "stuck" and therefore I will reluctantly use it.

A third camp, **the climate skeptics**, consists of those who are not sure that anthropogenic GHG emissions are the cause of any apparent global warming or even that any warming is occurring at all. This camp is therefore hesitant to side with either of the other two camps and is concerned more with the economic impact of policies enacted to reduce human-caused GHG emissions. Further, this camp is skeptical that any such policy changes will produce any significant reduction in the warming of the planet.

These camps have become known by the name given each by the other two. The first camp is known as the "**alarmists,**" camp two the "**deniers**" and camp three the "**skeptics.**" For the sake of clarity (implying no lack of respect for any) I will use those names throughout the remainder of this book.

In this chapter we will look at the Earth's temperature changes through time and how those changes are measured. In so doing we must recognize the great mysterious complexity of our Earth's atmosphere and climate.

There are, of course, many factors that influence global temperature and climate changes. Quite naturally any such discussion must start with our Sun, with its radiation and solar cycles, without which human habitation of our Earth would not be possible. This makes it the primary and most powerful factor driving temperature and climate. Others factors include, but are not limited to, the Earth's atmosphere, core, ocean currents, deforestation, reforestation, and volcanic activity. How these factors drive climate and global temperature is, for sure, a complex study requiring the understanding and ability to explain the interaction of so many factors and how each interacts with the others to produce the environment of the only habitable planet in our solar system. Any attempt to build models to predict changes in climate and draw any conclusions from the factors causing change must be inclusive of these many factors and also the degree to which they interact with each other. We must keep this in mind in our decision making process when enacting environmental policy changes.

A Brief History of Ice Ages and Warming

Long before the onset of the Industrial Revolution which began substantial development and use of the internal combustion engine, global warming and cooling had a constant history of occurrence. The Earth is currently warming after coming out of its last **Ice Age** 18,000 years ago as warming began and the Earth emerged from the **Pleistocene Ice Age.**[1] At that time much of the northern hemisphere was buried beneath glacier ice. The Earth's climate and biosphere have been dominated by ice ages and glaciers for the past several billion years, always in transition from warming out of them and then cooling back into the next glacier period.

We live now in a much more hospitable environment than in the deep freeze of those glacial periods. The relatively warm period we live in now is called an **interglacial period**. These periods tend to last about 15,000-20,000 years before returning back to those colder ice age climates and temperatures. Currently we are much closer to the end of this interglacial warming period than at its beginning.[2] Our concern is that anthropogenic GHG emissions are accelerating a current warming trend and that may cause an irreversible warming trend. To adequately assess the current warming trend of our planet requires we understand what caused the sudden warming 18,000 years ago. If we cannot, any measures taken to understand and limit this current perceived anthropogenic warming most definitely will be ineffective. I am not suggesting that since anthropogenic causes were not a factor in the warming 18,000 years that they indeed are not now. This is a subject scientists must address.

I think it is interesting to note and to present a brief review of some of the changes that have occurred in the Earth's environment since warming up from our last ice age 18,000 years ago[3]:

1. 15,000 years ago the earth had warmed sufficiently to bring to an end the advance of glaciers, and sea levels worldwide began to rise.
2. 8,000 years ago the land bridge across the Bering Strait was flooded, cutting off the passage of men and animals to North America from Asia.
3. Earth's temperature has increased approximately 16 degrees F and sea levels have risen a total of 300 feet. Forests have returned where once there was only ice.

Our civilization and modern society developed entirely during this recent interglacial period.

A Brief History of CO_2 Concentrations

Historically known through **proxies**, there has been much more CO_2 in our atmosphere than exists today. During the Jurassic Period (200 million years ago), average CO_2 concentrations were about 1800ppm or about 4.7 times higher than today's concentration which has reached 400ppm.[4] The highest concentrations of CO_2 during all of the Paleozoic Era occurred during the **Cambrian Period**, nearly **7000ppm** — about eighteen times higher than today.[2] Since this data is based on proxy and estimated data, its absolute accuracy (though informative) is, of course, questionable.

It is important to note from this particular historical record referenced above that temperatures 450 million years ago were about the same as they are today while at the same time CO_2 concentrations then were nearly twelve times higher than today. Though these numbers derived from proxy and estimated historical data may be lacking in absolute accuracy, we may cautiously use them for at least their comparative significance.

What the Past Tells Us About Temperature

We all know from our daily lives that the temperature of the Earth always changes; from hour to hour, day to day, year to year, and throughout the vast historical time periods or **ages** of the planet's existence. With the development of modern temperature-measuring instruments, temperature can be more precisely measured, creating more accurate historical temperature records for us to analyze. Even without these modernized instruments invented to measure temperature, we know (by means to be discussed below) that throughout the history of our planet, the Earth's temperature has warmed and cooled between ice ages during periods lasting for millions of years. Our concern now is that the temperatures in the last 150 years have been observed to be heating up more rapidly; about 0.8 degrees C.[5] Is this significant? And if so, is it being driven by human causes used to power the "**Industrial Revolution**?"

Paleoclimatology is the study of changes in climate taken on the scale of the entire history of Earth. It uses a variety of **climate proxy** methods to obtain preserved data to determine the Earth's climate history, to include methods of determining the Earth's temperature history. This reconstruction of historical data is critical in building models to predict future temperature changes. These modeling correlations can be made to show how isolating and changing factors, such as GHG concentrations, or other factors compare with historic temperature data, in analyzing their possible impacts on global temperatures. Predicting the future through the use of modeling is not a precise science. Scientists must know all factors involved and how they relate to each other to accurately predict future changes. Due to the complicated, and for sure not totally understood forces of nature affecting our Earth and atmosphere, these factors and how they impact temperature in relation to each other is not well known. It is, however, an important start in the prediction process, knowing all the while, its imprecision.

With relative confidence we know from Paleoclimatology that throughout the 4.5 billion year history of our planet its temperatures have varied between warming and cooling periods with each transition lasting for tens to hundreds of millions years. During the warm periods the polar regions were completely free of ice with forests extending all the way into those polar regions; and during the cold periods ice sheets and snow covered the entire surface of the planet.[6]

How do we know this? We know this from those climate proxies. Here are examples of a few good sources of this preserved data revealed by some of those proxies[7]:

—Mineral deposits in deep sea beds. Over time, dissolved shells of microscopic marine organisms create layers of chalk and limestone on sea beds. Analyzing the ratio of **oxygen-18** (a rare isotope) to **oxygen-16** (the common form) indicates whether the shells were formed during glacial periods, when more of the light isotope evaporated and rained down, or during warm periods.

— Pollen grains trapped in terrestrial soils. Scientists use **radio carbon dating** to determine what types of plants lived in the sampled region at the time each layer was formed. Changes in vegetation reflect surface temperature changes.

— Chemical variations in coral reefs. Coral reefs grow very slowly over hundreds or thousands of years. Analyzing their chemical composition and determining the time at which variations in corals' makeup occurred allows scientists to create records of past ocean temperatures and climate cycles.

— Core samples from polar ice fields and high-altitude glaciers. The layers created in ice cores by individual years of snowfall, which alternate with dry-season deposits of pollen and dust; provide physical timelines of glacial cycles. Air bubbles in the ice can be analyzed to measure atmospheric CO_2 levels at the time the ice was laid down.

Understanding the geological past is important to today's climate researchers, because they can study changing GHG levels and attempt to understand any correlations to changing temperatures and incorporate them into their models used to predict future temperature estimates. The obvious question is "How do we know that increasing concentrations of GHGs are causing an increase in global temperature?" We have known for many years the impact GHGs have on trapping heat within our atmosphere. As discussed in chapter 3 we know why GHG molecules, which have at least three atoms, can absorb and emit heat better than oxygen (O_2) or nitrogen (N). This is extremely well understood. The measurement of the CO_2 molecule emission of energy was first discovered in 1863 by James Tyndall, a nineteenth century physicist, who first demonstrated that when CO_2 molecules are doubled in the atmosphere, 4 Watts of extra energy will be released (radiated) in every square meter of the planet every day.[8] That would warm the surface. This experiment has been performed thousands of times in laboratories all over the world with the same results. This is the radiative **forcing** used in the IPCC global warming modeling.

We know that humans are adding CO_2 molecules to the atmosphere at a rate of about two parts per million every year, and this rate has increased in recent decades, dramatically. We must be careful, however, when we look at graphs depicting this apparent dramatic increase. Here is a good example of how changes can be depicted and perspectives distorted.[9] In the graph on the right it looks like CO_2 concentration is dramatically increasing in the latter part of the twentieth century. This could be misleading in that it depicts seemingly huge concentration increases.

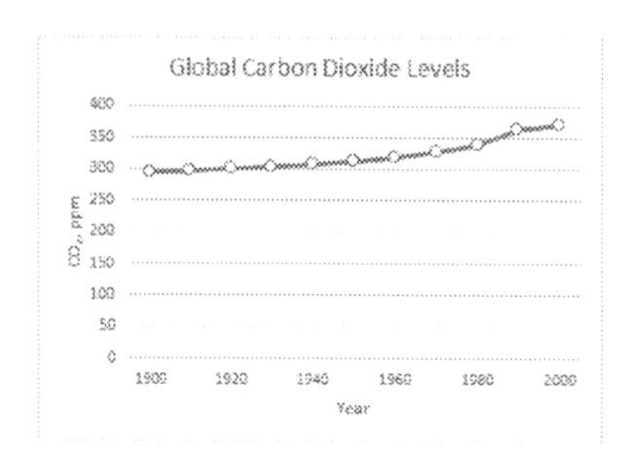

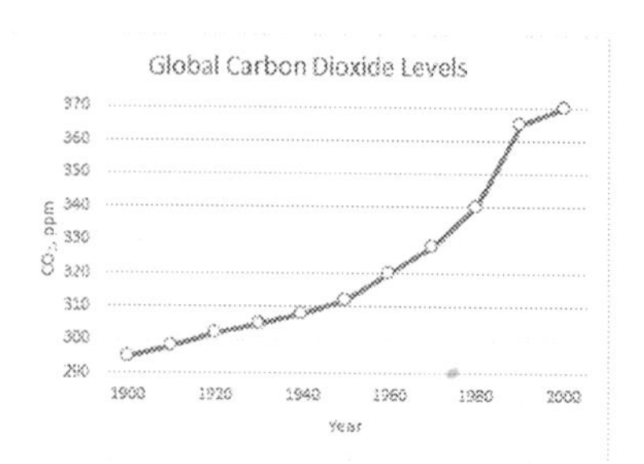

When the same data[10] is plotted with the y axis starting at zero as in the graph on the left, a far less dramatic slope of the same increase in CO_2 is shown for the same time period. In hard science journals, the graph on the right would be considered deceptive because the y-axis starts at 290 instead of zero. This misleads the reader into thinking that CO_2 concentrations have increased at a much greater rate.

The question we must ask ourselves is "to what degree does the atmosphere respond to this forcing of increasing CO_2 concentration?" Our climate system is chaotic and does not require us to upset its balance—it does that quite well on its own as Paleoclimatology has shown us. Further complicating this understanding is a study published by Dr. Manfred Mudelsee, a German physicist and founder of Climate Risk Analysis who measured the gas bubbles trapped in ice cores from Greenland and Antarctica. His study concluded the close correlation of changes in temperature and CO_2 concentration in such cores occur in close parallelism. "In detail, however, the changes in temperature precede their parallel changes in CO_2 by between 800 and 2000 years (Mudelsee 2001). This study would suggest that CO_2 cannot be the primary **forcing agent** for temperature change at the glacial-interglacial scale.

The two obvious concerns in formulating policies and solutions for policy makers are to first determine how our atmosphere responds proportionally to:

1. The forcing of anthropogenic GHG emissions,
2. Natural forcings, or
3. Some of both

Our government has spent a lot of money researching the first concern of anthropogenic causes of global warming and climate change, but the public almost certainly does not realize the government has virtually ignored the potential of natural causes of warming and climate change in government-funded research. I will deal with this insight in later chapters.

Understanding Global Temperature Change

The oceans dominate the long term aspects of climate change because they can hold so much heat for a long period of time. In response to forcings like the four Watts/meter^2 there is a **feedback**—the Earth pushes back when it gets too hot or cold, and the disagreement is the degree to which the Earth does that. The oceans can trap a lot of the planet's heat and slow Earth's surface heating. How the Earth reacts to these forcings and how it pushes back is called climate sensitivity. Those who believe anthropogenic influences on global warming have a major impact believe in a high degree of climate sensitivity. Those who are more skeptical of anthropogenic impacts consider natural causes such as changing solar radiation, as discussed earlier, the changing frequency of **El Ninos** vs **La Ninas,** and wind patterns caused by **Arctic Oscillation** are also significant and, therefore, believe the climate sensitivity to CO_2 is much lower.

We know from the science of paleoclimatology that climate has varied throughout the five billion year history of our planet.[11] The question and concern before us now is: "Is the observed temperature increase in the past 150 years or so a normal variation caused by natural causes or an increase accelerated due to a positive feedback forcing caused by human action?"

These early temperature estimates were made through a variety of indicators using the different oxygen **isotopes** (oxygen 18 and oxygen 16) found in the shells of animals as presented earlier. "The concentration of oxygen 18 in precipitation decreases with temperature as it condenses out first. The lighter oxygen 16 evaporates first. In colder time periods the shells contain more of the oxygen18 isotope and in warmer periods more of the 16 isotope.[12] This can then be correlated with the oxygen isotope ratios from ice cores, and sea floor sediment cores and temperature estimates can then be made. These measurements are made over huge time periods where fluctuations are diminished (flattened) by those large scales of time. When we talk about a few degree change in measured temperature over the past 150 years it will look much more amplified than if graphed over larger geological time periods. Don't be misled by slopes on graphs.

A Brief History of the Science of Global Warming and Cooling

Anyone trying to stay up to date on the issues of global warming is bombarded with conflicting statements from those professing different "scientific" viewpoints. "There is a 97 percent consensus among scientists…" "the 97 percent is deceiving;" "the Earth is warming;" "the Earth is cooling;" "the Earth's temperature is stable;" "the ocean levels are rising;" "the ocean levels are not rising at any greater rate;" "the arctic ice melt is unprecedented;" "the arctic ice is in a normal state of change;" …and on and on the debate continues. I will deal with these apparent conflicting positions later, but let's first focus for a moment on the history of global temperature variations.

By drilling **ice cores** deep within the icepacks of Greenland and Antarctica, Scientists can study Earth's climate as far back as 800,000 years. Trapped within those cores are air samples providing information on air temperature and CO_2 levels. In the 1990s the Vostok (a location near the South Pole) ice core showed temperature and CO_2 moving in the same direction at the same time. It made sense to assume from this graph[13] that CO_2 did influence temperature:

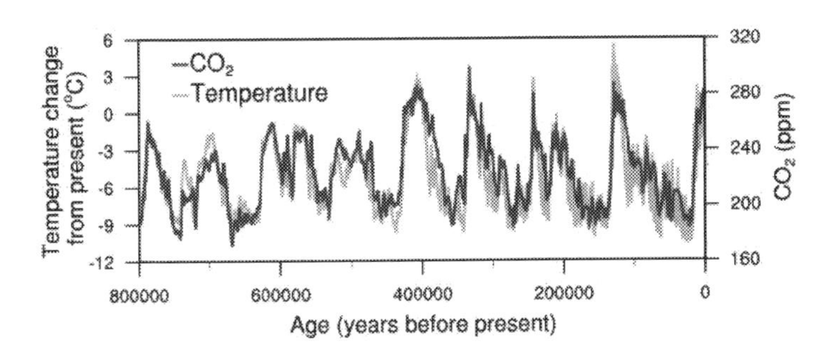

Then in 2003 new data was introduced when the time frame of the graph was expanded to show that CO_2 concentrations actually lagged behind temperature change leading to the theory that it was actually temperature that drove CO_2 levels. This point is argued by scientists like NASA's **James Hansen** claiming that initial changes in temperature during this period should be expected, explained by changes in the Earth's orbit around the sun warming the oceans causing increased amounts of CO_2 to be released. That, however, would reveal global warming possibly driven by natural variation?

The graph[14] below depicts the rise and fall of CO_2 lagging behind temperature by about 800 years.

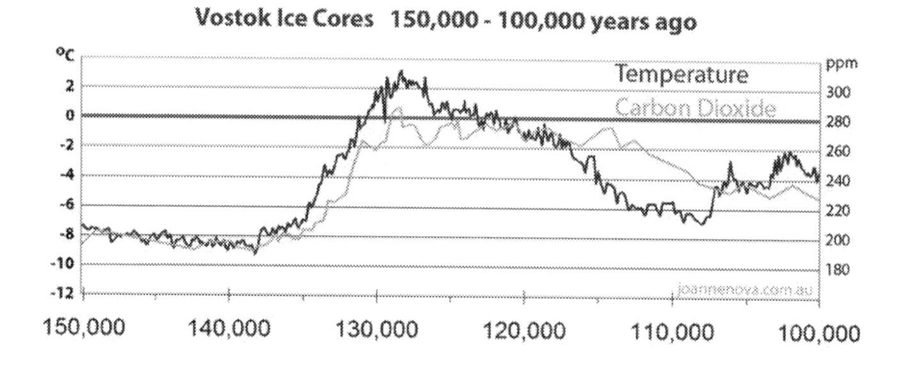

Graph Courtesy of Joanne Nova
Joannenova.com.au

This finding led to the theory that if CO_2 was the major driver of temperature, why would temperatures not rise indefinitely in a runaway greenhouse effect? "This hasn't happened in 500 million years, so either a mystery factor stops the runaway greenhouse effect, or CO_2 is a minor force, and the models are missing the dominant driver."[15] This study suggests the ice cores don't prove or disprove what has caused past warming or cooling but introduces a simple explanation that when temperatures rise, more CO_2 enters the atmosphere as the oceans warm and release more CO_2 into the atmosphere. If this study cannot be disproved, something else must be causing the warming.[16]

Modeling Prediction

This graph, constructed from latest IPCC report[17] (to be presented in chapter 6) depicts the last two decades of recorded temperature data plotted with the predicted future projections based on **Representative Concentration Pathways** (RCPs) derived from modeling inputs that allow for uncertain outcomes.

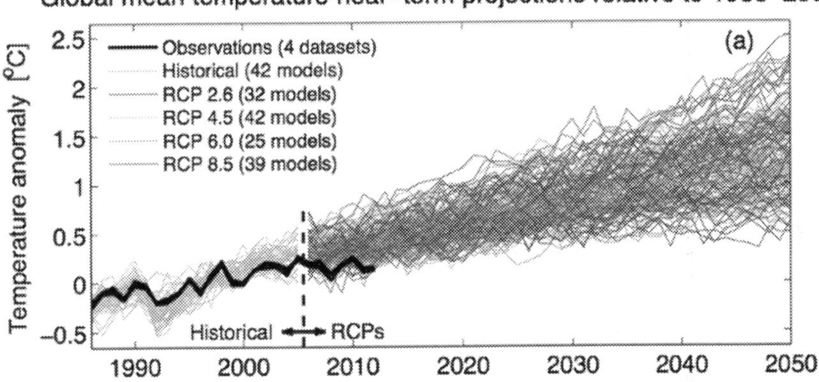

As can be seen the observed temperatures since 2000 have leveled off and are below the projections of the models used by the IPCC.

It is true that since 1998 was a hot year, using it as a starting point for any graph would "flatten" subsequent years' temperatures, but a starting point anywhere before 1998 would still show this slowing of warming. Using a starting point of 1990 or 2000 would show a more upward trend, but such is how graphed data can be presented to try to make a point, as previously discussed.

The pesky issue in this temperature observation since 1998 is the rapid increases in the amount of GHG emissions that have occurred during that time with little evidence of increase in warming. Andrew Weaver, a University of Victoria climate scientist who is also a Green Party member of the British Columbia parliament responds: "The heat is not missing. The heat is there. The heat is in the ocean."[18]

If it is, and certainly it may be, could this not be a feedback or even a pushback in this complex global phenomenon of the climate of this planet to maintain a means to regulate global temperature for which current modeling does not account?

How is the average temperature of the Earth determined?

In attempting to determine the "average" surface temperature of the Earth, we only add to the difficulties of understanding the problem of documenting the occurrence or non occurrence of global warming. Much attention has been directed toward placements of the "thermometers" used to take surface temperatures. For example if a thermometer is placed in the city at a location surrounded by concrete and high heat reflective surfaces it will record a different temperature (and more rapidly rising temperature as construction in those area continue) than, for instance, a thermometer in that same city placed in a location such as the city park usually surrounded by less heat reflective surfaces. For those of us living in coastal areas, we know when the weatherman predicts an overnight freeze; it is very likely that those living closer to the coast may not see that freeze because of the influence of oceans on the air mass closer to the shoreline. Scientific modeling is suppose to account for thermometer placement, but how accurate those placement modeling factors are should be suspect. At any rate, from all these globally-placed thermometers we compile records over time and can determine temperature trends. That is where we find ourselves now, noticing a general temperature rise in the last 150 years or so, and then a "leveling off" of temperatures in the last two decades. Recently, NASA concluded that globally averaged temperatures in 2015 made it the hottest year on record "shattering" the previous 2014 record by 0.13 Celsius,[19] but also concluded this claim was within the margin of error for these calculations. Also it must be pointed out that a warming **El Niño** was in effect for most of 2015. To be sure, from the perspective of most of us, these temperature increases are extremely small leading to questions on how accurate and significant can these small increases be. A big part of the problem in temperature data gathering is in trying to obtain statistically valid conclusions when raw data must be "manipulated" using algorithms that consider the varied spacing of temperature stations around the globe and urban heating effects that could skew the conclusions if left unaccounted for. Although I'm not accusing anyone of manipulating data to produce a desired outcome (that subject will be discussed in chapter 11) the fact that that temperature data is manipulated[20] to account for varying placement of temperature gather equipment lends itself to that suspicion. Also we recognize that analog temperature measurements from the previous two centuries are less accurate than current state-of-the-art digital measurements which, of course would lead to larger deviation errors using differing measurement technologies. The important thing to remember here is that we are talking about a very small temperature increase in the last 150 years—0.8 degrees Celsius. Using the word "shattering" to describe new temperature rises might be considered a little too dramatic in stating the case for global warming.

The Hockey Stick: The Most Controversial Chart in Science

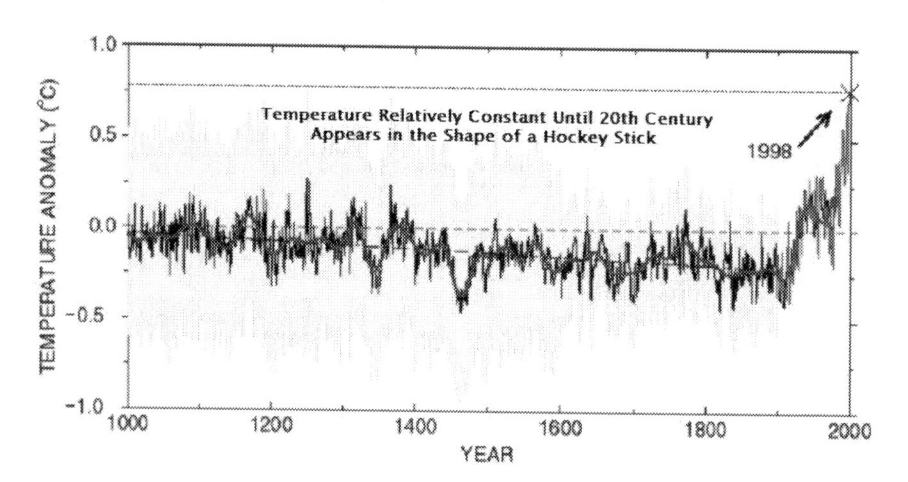

In 1998 Michael Mann and two colleagues, Raymond Bradley and Malcolm Hughes introduced a graph[21] that sought to reconstruct the planet's past temperatures going back half a millennium before the era of thermometers, attempting to show how comparatively accelerated recent warming has been.

This graph became an iconic symbol that suggested humanity's contribution has in fact accelerated global warming. Mann's finding (from above referenced graph): Recent northern hemisphere temperatures had been "warmer than any other year since (at least) AD 1400." The graph depicting this result looked much like a hockey stick: because its long handle corresponds to 900 years (from 1000 to 1900) of little temperature variation, and its blade represents 100 years (1900 to 1999) of rapid temperature rise.

The graph was featured in the summary of a crucial IPCC report (its Third Assessment Report—these assessments will be presented in chapter 6) by a United Nations climate panel from 2001 and appears to show a rapid temperature rise (1 degree C) around the turn of the twentieth Century. Graphically this displays the basic assertion that the planet's recent warming is unprecedented over at least the last millennium. The darker shaded lines on the graph are actual temperature recordings from thermometers and the lighter shaded lines with gray shading depicts the confidence bands of temperature reconstruction using proxy temperature recording methods. The "hockey stick" graphic gives the appearance that left to its own devices, nature displays very little in the way of temperature variation, but that during the past century, humans have come along and influenced a dramatic change in what appears to be a pattern of normal variation. It is thus the "perfect representation" of the greenhouse gas message that humans have caused this rapid increase in temperature which the IPCC gave much publicity.

Since Michael Mann debuted this graph in the Geophysical Research Letters in 1999, he has been defending it from scientists skeptical of his findings.

To many skeptical scientists, including Harvard scientists Willie Soon and Sallie Baliunas, the shape of the "hockey stick" did not corroborate the existing knowledge of the climate of the past millennium.[22] They asked: "Where was the **Little Ice Age**" the well-documented cold period lasting from about the sixteenth to the nineteenth century? And "where was the **Medieval Warm Period**," a relatively warmer period extending from about eleventh to the thirteenth century (when Viking exploration resulted in their farming settlements in Greenland, for example)? By not showing any indication that these climate episodes existed, the "hockey stick" appeared to present a completely new picture of the climate of the past 1,000 years and was used extensively by the IPCCs third assessment in 2001. However, this was not the same story that is being told in countless weather and climate textbooks used in classrooms around the world. In fact, as we shall see in chapter 6, the IPCC in its fourth assessment in 2007, changed its graph (shown below) on temperature variation during this time period to reflect the Medieval Warm Period and the Little Ice Age using data by Moberg et al. (2005) on temperature reconstruction in the northern hemisphere.

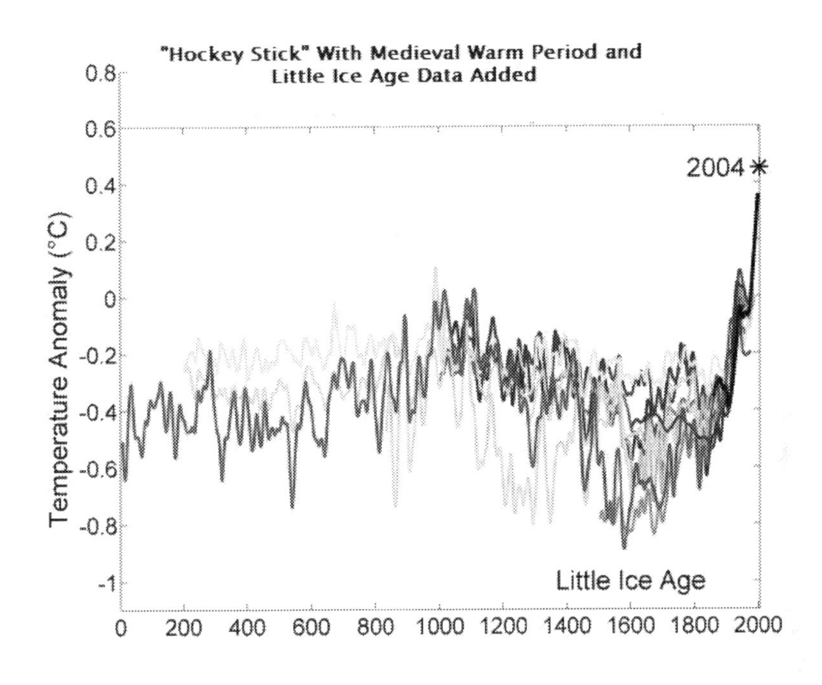

Then came the work by Steven McIntyre (a mineral consultant) and Ross McKitrick (an economist) and the author of "Taken By Storm," which attempted to reproduce the "hockey stick" using the data and procedures described by Mann and his colleagues in their 1998 *Nature* publication.[23] McKitrick and McIntyre decided out of personal interest to audit on their own the data used to create the "hockey stick" graph and see if they could recreate it starting from scratch. For years they worked on this recreation facing many road blocks from the creators of the hockey stick in gaining access to their data.[24] When Mann and the other authors of this study refused to release the data, Congress, concerned with policy decision-making, became involved and established the Wegman Committee under Edward Wegman, a professor of mathematics and statistics. This investigative committee requested and obtained the data. His report, submitted in the summer of 2006, stated in brief: *"Mann et al., misused certain statistical methods in their studies, which inappropriately produce hockey stick shapes in the temperature history. [the] analysis concludes that Mann's work cannot support the claim that the1990s were the warmest decade of the millennium."*[25]

McKitrick noted: "While no individual mistake was likely sufficient enough in and of itself to throw into question the "hockey stick," taken together, the list of errors indicate a certain lack of rigor and attention to detail by the "hockey stick's" creators. (Shipley 2011)

A third dissenting voice was that of Jan Esper and colleagues in 2004. Esper's expertise is in climate reconstructions based on tree-ring records (the primary type of proxy data relied upon by Mann et al. in creating the "hockey stick"). Esper asserts as the tree itself ages, the widths of the annual rings that it produces changes—even absent any climatic variations. He maintains this growth trend needs to be taken into account when attempting to interpret any climate data contained in the tree-ring records.[26] When this growth trend is ignored, he points out, long-term climate trends are lost as well and this could be one likely reason why the handle of the "hockey stick" is so flat. Using an alternative technique that attempted to preserve as much of the information about long-term climate variations as possible from historical tree-ring records, Esper and colleagues derived their own annual Northern Hemisphere temperature reconstruction. (Shipley 2011) The result is a 1,000 year temperature in which the Little Ice Age and the Medieval Warming Period are much more pronounced than the "hockey stick" reconstruction—evidence, in his opinion, that the "hockey stick" underestimates the true level of natural climate variations.

The Controversy Over Mr. Mann's Graph Continues

Despite these studies the controversy continues with Michael Mann and his supporters vigorously defending the "hockey stick" depiction and its significance. Another National Academy of Science study in 2006" vindicated the hockey stick as good science noting:

"The basic conclusion of Mann et al. (1998, 1999) was that the late twentieth century warmth in the Northern Hemisphere was unprecedented during at least the last 1,000 years. This conclusion has subsequently been supported by an array of evidence that includes both additional large-scale surface temperature reconstructions and pronounced changes in a variety of local proxy indicators, such as melting on ice caps and the retreat of glaciers around the world."[27]

The question remains throughout the global warming debate: "Who is right, Michael Mann or those skeptical of his work?" The controversy continues. The science does not appear to be settled.

Weather versus Climate

In studying temperature variation over time, we must guard against confusing weather with climate. Those who would point to current weather "events" as evidence of a change in climate are myopic. For example, the following image with data taken from Applied Climate Information Systems[28] illustrates this myopia. In March 2012, those who align themselves with the "warmist" view of climate change pointed to that month as "proof" to many of global warming.

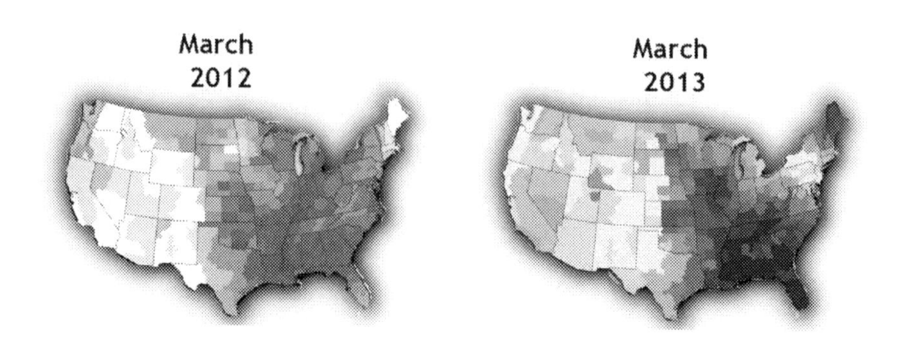

A heat wave crossed the nation causing very unusual weather patterns. In my home state of Minnesota the snow was gone and for the first time in many "old timer's" memories the ice was gone from most of its many lakes. Just one year later in March 2013, the opposite occurred with very cold temperatures, causing record cold and snowfall amounts. The "old timers" again proclaimed they'd never seen anything like that and many skeptics proclaimed this was proof that global warming is a myth. Both viewpoints are flawed with shortsighted assessments. This is a classic example of not understanding the difference between weather and climate. A further flaw of both these viewpoints is "where is the global perspective?" These are weather maps of the United States, not the planet. We need to keep our perspective "global" when we make global warming/cooling judgments and policy decisions.

Then in February 2014, John Podesta, one of President Obama's top climate advisers, presented the following two contrasting satellite images of California's snowpack. The one on the left was taken in January 2013. The one on the right was taken a year later and it emphasizes how global warming is changing the environmental conditions in the United States.

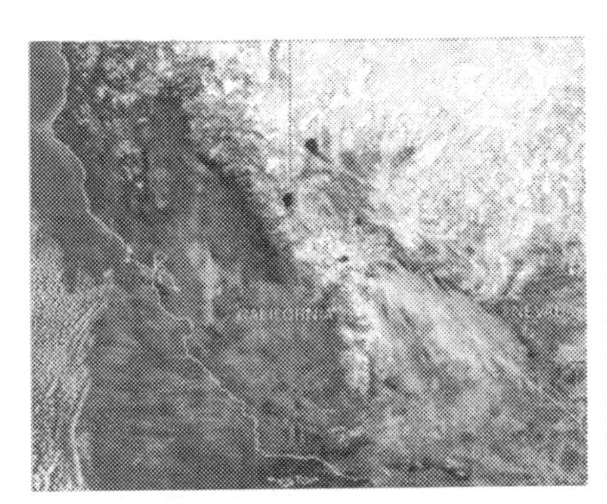
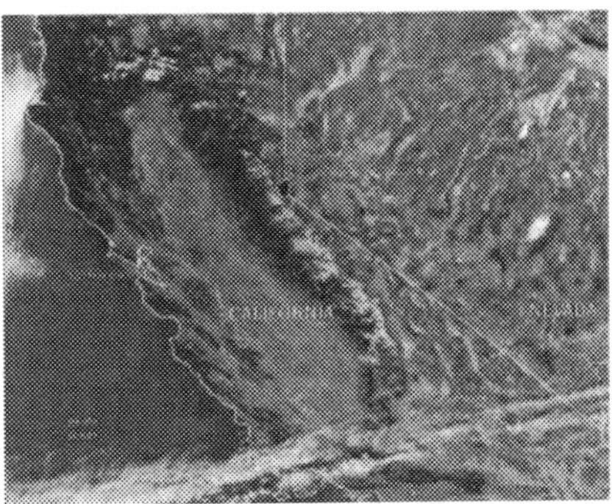

Satellite image courtesy of NOAA

President Obama then passed on Podesta's warning to a meeting later that day of western Governors saying political leaders had "no choice but to cope with global warming's impact."[29]

We must guard against the notion that year to year weather variability is "climate" and that global warming is to blame for these variable and seemingly more extreme weather occurrences. Such notions in my opinion are narrow-minded. To this point Scientists at the National Oceanic and Atmospheric Administration published a study in 2014 stating *"there is no discernible connection between global warming and 2013 extreme weather events such as the California drought, Colorado floods, the UK's exceptionally cold spring, a South Dakota blizzard, Central Europe floods, a northwestern*

Europe cyclone, and exceptional snowfall in Europe's Pyrenees Mountains."[30]

Modeling

The Earth's climate system is extremely complex, so much so that to understand it and to predict any changes, especially in the long term, is very difficult to do. A scientific tool, yet in its infancy, used to assist in such prediction is **global climate modeling** (GCM). To accurately predict future climate, the models used to make these predictions must include all of the factors affecting climate as well as the scale to which each interact with one another to impact the broad range of environmental conditions. This is daunting if not impossible task for any model to accomplish since all those factors and their relative influence on the Earth climate system cannot be wholly known or understood. Still they can be useful. The following pictorial represents some of the inputs of the **Community Earth System Model** (CESM) used by scientists at the North Carolina University for Atmospheric Research and considered to be one of the most sophisticated and powerful models to simulate the many processes in our climate system.[31] How these factors interact with each other is critical in constructing accurate predictions from any model. That science is beyond the scope of this book, but it is important to point out its limitations.

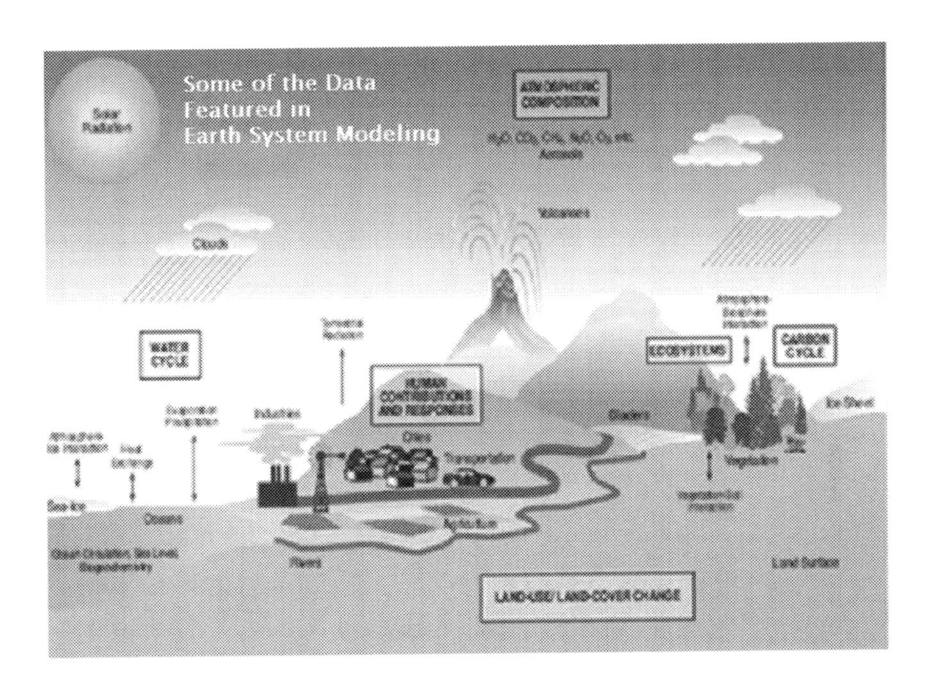

The IPCC models have been criticized for overlooking much of the effects of solar radiation and the associated solar cycles because they are not well understood.[32] Yet it cannot be overstated that the dominant energy source of our planet is its Sun and the manner in which solar energy is distributed by both radiation and convection as what drives the climate system. That many of these cycles and possible associated mechanisms by which the Sun influences Earth's temperature and climate are poorly understood is no justification for ignoring them (Carter 2017).

I will present the importance this modeling has on policy decisions in chapter 11.

The Hottest Year in Modern History

At the beginning of this chapter I quoted NASA Scientist James Hansen and Vice President Gore as saying 1998 was the "hottest year in modern history." So are they correct? If not, what was the hottest year recorded in modern history? I am sure that 1998 and those in the running for this environmental claim will be argued at length and will be a "moving target," always being revaluated by scientists. Part of the conflicting claims for this seemingly important "fact" is because that actual number holds such a slim advantage over the others. We're not talking about one particular year being 5 degrees hotter than the "next hottest," but rather in the order of "hundredths" (.01 degree) of a degree, surely within the margin of error of those data sets.

Having proposed this question, NASA has now backtracked on its 1998 warmest year claim and admitted that those claims of James Hansen were based on "mathematical errors." NASA now claims they have corrected that error, and 1934 is now known as the warmest year on record, 1998 second, with 1921 the third warmest year instead of 2006 as was also previously claimed.[33] Again we're talking about hundredths of a degree.

NASA has had to re evaluate their figures because for the same reason modeling is flawed when all the factors are not known and therefore not included, the temperature figures used were distorted by the urban ground (**heat islands**) where most of these measurements took place. The ground is warmed considerably by human activities and cannot accurately represent atmospheric conditions.

Conclusion

Historical evidence survives that the Earth's climate and temperature fluctuate continually over time. Evidence is also clear that in the last century the global temperatures have risen about 0.8 degrees C until about 1998 and then has remained mostly constant. The matter for scientists to solve is "what causes these fluctuations in both climate and temperature?" And the even bigger question, "What role does human activity play in those fluctuations?" We know CO_2, among others, is a GHG that traps heat in our atmosphere and that its concentrations, also fluctuate and have done so long before humans inhabited the Earth. We have seen studies that show global temperatures rise with increased CO_2 concentration, and studies that show CO_2 rise also follows a rise in global temperature. We have witnessed in the last two decades a rapid rise in CO_2 concentration with no significant rise in global temperatures that were predicted by modeling. We have seen that the actual temperature gathering methods have changed dramatically over the last two centuries from the very "crude" (for their time), to the very sophisticated satellite measuring methods that in and of itself affects measurement analysis. We also know that our complex atmosphere and climate system respond to an even more complex feedback system that science still cannot explain.

As the argument goes for some, while the atmosphere is getting warmer and since humans have been putting more and more CO_2 into the atmosphere wouldn't it be logical to assume the two are related. But we know from science that correlation does not always mean causation. Could there be any alternative explanations for recent warming that would diminish the human contribution to this warming? Yes there are. Have scientists ruled out any of these explanations? No they have not. Before global warming became the hot research topic it has recently become, natural cycles of global warming and cooling were the rule not the exception in most of the evidence published. According to Dr Roy Spencer, former NASA climatologist, a good example of these natural cycles of warming and cooling would be the occurrence of the Medieval Warm period that appeared around 1000 AD. Evidence from proxies suggests it was warmer then than now. Dr Spencer continues the point that modern science has yet to understand these natural cycles.[34] Science cannot study what it does not have the data to study. For example the recent warming trend noticed that has occurred since the late 1970s might have been mostly caused by a global average decrease in cloudiness due to a natural internal climate cycle. The data to document such a change has only existed since the year 2000.[35] If it turns out that these natural climate cycles are the cause of global temperature change, that would lead us to believe that we have a very robust climate system not nearly so sensitive to added CO_2 in the atmosphere. The observation that there has been a pause for the last two decades in warming, suggests that scientists may have overestimated the role of CO_2 in climate change.

CHAPTER FOUR

EVOLUTION OF ENERGY USE

"If you want to find the secrets of the universe, think in terms of energy, frequency and vibration."
—Nikola Tesla

Introduction

From mankind's first reaction to the pleasant warmth of the sun and the realization that he could recreate this comfort by burning the seemingly limitless supply of heat producing resources surrounding him, he quickly learned to adapt to his environment and venture into harsher ones, while improving the quality of his life. This desire of us humans to improve the quality of our lives is the driving force behind why we have a seemingly insatiable appetite to harness and consume energy.

Certainly one of the first controllable energy sources discovered by man was fire. When mankind first learned to harness, use, and control fire, he began to emit human or anthropogenic-caused emissions of greenhouse gases (GHGs) and other gases and particulates into the atmosphere. No other species on Earth ever learned to "make heat." As humans gained knowledge of what could be used to generate heat by fire, he branched out from wood and coal, to petroleum. Along the way the energy sources of the sun, wind, hydro, hydroelectric, solar, and nuclear were incrementally added as energy sources as well.

In the eighteenth century with the development of the steam engine, massive amounts of energy were capable of being produced enabling accelerated growth in the production of mechanical energy from the process of burning carbon resources, harnessing water power, and transitioning from wood to coal and oil. This mechanical energy could then be used to run the ever-developing machinery of the Industrial Revolution, to do work, and open up so many new possibilities to improve the quality of life influencing virtually every aspect of daily human life. The steam engine was probably the first major invention that allowed the Industrial Revolution to take place. Steam engines gobbled up hundreds of millions of years' worth of stored energy trapped in both fossil fuels (coal) and trees to heat water in large boilers and produce the steam that drove the machinery of the Industrial Age. As a result mankind began to have a much greater impact on the environment of our planet. Prior to this time the footprint of man was considered nearly inconspicuous. With the Industrial Age, began the obvious and persistent increase in the "footprint" being left by humans. From air quality to permanent buildings, some rising hundreds of feet above ground, to the development of permanent roads, rail lines, mines, forestry harvests, and massive cities of sprawling concrete and steel, man began to leave his ever-increasing mark on this planet.

The worldwide energy harvest increased fivefold in the nineteenth century. It rose another sixteen-fold in the twentieth century with oil. No other century, or any other millennium in human history can compare with the twentieth century for its growth in energy use. In the 100 centuries between the dawn of agriculture and 1900, people used only about two-thirds as much energy as in the twentieth century.[1]

Unfortunately for most of the entire undeveloped world, this industrial revolution either did not take place at all, or proceeds forward at a very slow pace. A billion people on this planet are still without electricity and must use wood and animal dung for heating and cooking. For these people life has remained relatively unhealthy and short. Everything humans do requires energy and while the developed world is now moving away from those early forms of energy so heavily dependent on fossil fuels, the undeveloped world for economic reasons remains largely "trapped" in those early energy forms that remain available and affordable having few other affordable energy options. This creates an obvious problem in dealing with reducing GHGs as well as air and water pollution on a worldwide scale. The cleanest and healthiest environments exist in the industrialized world. This fact weighs heavily in the politics of enacting environmental change and will be discussed in chapter 12.

Until recently the incentive to move to more "greener" energy sources was driven by the belief that fossil fuel resources would sooner, rather than later, be exhausted and become more expensive. However, technology has produced the means to economically find and extract fossil fuel energy well into our future while making it more economical than the expensive and heavily taxpayer-subsidized green energy sources. This technology that has become a major contribution to the evolution of energy is the process of **fracking** which makes available vast new supplies of fossil fuel energy to almost every part of the Earth. Fracking is giving this country and others the possibility of acquiring and maintaining energy independence. This process, that has been used for the last fifty-plus years, requires drilling down into the abundant oil-laden shale rock beneath the Earth's surface and pumping under high pressure a water-sand-chemical mixture that releases oil, but primarily the more clean-burning natural gas from the shale rock formations. The potential of this process to produce cheap and cleaner energy far into the future is huge. It is also important to note that this process and technology does not come without controversy. The process requires a solution 98 percent of which is water and sand and 2 percent chemicals, such as boron, acetic acid (the active part of vinegar), hydrochloric acid and sodium hypochlorite acid (the latter two both used in swimming pools). The process requires large amounts of water which can later be reclaimed. The risks taken with this fracking process are the potential contamination of ground water and aquifers. Many studies have been conducted on the potential hazards of the fracking process, some indicating contamination is occurring, and others showing the process safe (University of Cincinnati). [2] Most of these studies claim opposing studies as inaccurate or incomplete. The definitive study has not been published. As with all mining operations around the world, contamination is always a concern and requires the safety precautions of oversight.

Nuclear energy is possible only for the wealthiest of nations. It comes with controversy because of potential hazards, and disposal of its waste material, but remains the cleanest form of abundant energy. This energy source will be discussed further in a later chapter.

The next phase of the evolution in energy development is upon us now as our environmental concerns direct us toward the "greener" energy sources with the goal of creating a carbon-neutral world, using nuclear, hydro-electric, wind, and solar energy sources. The application and feasibility of using these energy sources to gain a more carbon-neutral energy base globally will also be discussed later.

CHAPTER FIVE

INTERNATIONAL ORGANIZATIONS RESEARCHING GLOBAL WARMING AND CLIMATE CHANGE

"When science is politicized it is no longer science."
—Willie Sune, Astrophysicist, Harvard-Smithsonian Center for Astrophysics

Introduction

The objective of this chapter is to present two of the major and often conflicting positions of scientists on the causes of global warming and climate change. This is not a chapter about consensus for there can be no scientific foundation in consensus; rather the opposite. It is about the basis of the dissimilar positions of two organizations that maintain two different "scientific" positions on the subject of global warming and climate change. Both claim they have science to support their conclusions. These two organizations are the United Nations' Intergovernmental Panel on Climate Change (IPCC) and the Nongovernmental International Panel on Climate Change (NIPCC).

Both of these groups were organized to deal with the science and social issues of global warming and climate change and to provide policy advice. Both organizations claim science is on their side, yet these two organizations are in almost every respect opposed in their findings and generally show a lack of professional respect for each other. The IPCC claims NIPCC affiliation with the Heartland Institute, lacks scientific credibility and the NIPCC claims the IPCC's affiliation with governments give its findings a lack of scientific credibility as well. How can their findings be so different? This chapter will make the case that the disconnect between these two scientific communities demonstrate that the science of why the Earth is warming and causing subsequent changes in our climate is incomplete and has a long way to go in understanding the influence humans have on perceived changes in the Earth's environment.

I said at the outset of this book that my objective was to keep politics out of this book. In presenting these two organizations, this task may prove difficult, but any analysis of these two organizations would be remiss if it did not account for the political dynamics that influence and make them up and the reports they publish. The purpose of this chapter is to present and to inform the reader the views and scientific perspective of these two groups. I will strive for objectivity, in presenting this dynamic, and save the politics for a later chapter. I would again ask the reader to leave any preconceived ideas "at the door."

The IPCC

The IPCC was started in 1988 by the United Nations as the policy advocacy arm of the United Nations to address concerns that certain human activities could change global climate patterns and threaten present and future generations with potentially severe economic and social consequences. By direction, the IPCC is charged not to conduct research but to make comprehensive scientific assessments every five years of current scientific, technical, and socio-economic information gathered from worldwide research reports about the risk of climate change caused by human activity. Further it is directed to estimate potential environmental and socio-economic consequences, and to assess possible options for adapting to these consequences.[1] The IPCC defines climate change as the "significant changes in global temperature, precipitation, wind patterns and other measures of climate that occur over several decades or longer." The focus of the IPCC has been to predict future climate change based on computer models developed from knowledge gained over the last 150 years of instrumented climate observations, as well as proxy measurements such as tree ring analysis. Though the main objective of this organization is to analyze the human influence on global warming and climate change, the IPCC also recognizes the need for additional research and scientific study into "all sources and causes of climate change,"[2] though as we shall learn, few resources are spent in this latter research.

The IPCC is an organization of government appointed officials, currently representing 195 countries that are members of the United Nations, and funded by the UN through a trust fund. Since 1988, the IPCC has published five sets of "assessment reports" (AR) on the state of climate science based on the research and reports of hundreds of scientists from across the world. These assessments are made by government bureaucrats, not the research scientists who author those assessments. They publish their assessments within three working groups: Working Group I (WG1), the physical science group; Working Goup II (WG2), the impacts, adaptation, and vulnerability group; and Working Group III (WG3), the mitigation of climate change group. These authors are not paid for the time they commit to the IPCC, rather they belong to other institutions from which they earn their incomes. Some of these institutions are well-known like the National Oceanographic and Atmospheric Association (NOAA) and the National Center for Scientific Research to lesser-known scientific institutes from around the world mostly based in Universities.

The NIPCC

The NIPCC is an international panel of nonprofit, nongovernment scientists and scholars with no formal relationship with any government or government agency. It is a project of three independent nonprofit organizations: Science and Environmental Policy Project (SEPP), Center for the Study of Carbon Dioxide and Global Change, and The Heartland Institute., who have come together to present a comprehensive, authoritative, and realistic assessment of the science and economics of global warming. Contributions to all three organizations support the project. Although a lesser known organization than the IPCC, it consists of an international network of independent and equally distinguished scientists from fifteen countries. The NIPCC is involved in tracking and analyzing climate change, and is able to offer an independent "second opinion" of the evidence reviewed, or not reviewed, by the IPCC on the issue of global warming and climate change. Because this organization is not a government agency, and because its members are not predisposed to believe climate change is caused by human greenhouse gas emissions, the NIPCC can claim they are free from any political pressure or influence in producing any conclusions and can make policy recommendations based on purely the evidence of science.

Dr. Fred Sanger, an Austrian-born American physicist and professor emeritus of environmental science at the University of Virginia, along with the Science and Environmental Policy Project (SEPP) launched the NIPCC in 2003 in anticipation of the release of the IPCC's Fourth Assessment (AR4). The panel's stated purpose is to objectively analyze and interpret data and facts without conforming to any specific agenda.[3] This organizational structure and purpose stand in contrast to those of the United Nations' Intergovernmental Panel on Climate Change (IPCC), which is government-sponsored, and predisposed to believing that climate change is anthropogenic in cause and a problem in need of a U.N. solution. Subsequently the NIPCC partnered with

The Heartland Institute to highlight what they found as deficiencies within the IPCC's AR4.

The NIPCC has taken the lead among the skeptics on prevalent claims regarding the causes of climate change. In 2009, the NIPCC released its first report titled *Climate Change Reconsidered: The Report of the Nongovernmental International Panel on Climate Change.* In September 2013, NIPCC released its second report: *Climate Change Reconsidered II: Physical Science,* bringing their research up to date with the IPCC. These two reports contained two 1000 page volumes prepared by a large number of scientists and has been peer reviewed by nearly 6,000 studies.[4] In its May 26, 2012 issue *The Economist* called the Heartland Institute "..the world's most prominent think tank promoting skepticism about man-made climate change," and in May 2012 the New York Times called them "The primary American organization pushing climate change skepticism."[5]

It should come as no surprise that the authors (including scientists) of the IPCC and the scientists in NIPCC have an adversarial relationship. The NIPCC is highly criticized by supporters of the IPCC for what they claim presents a lack of scientific evidence by a mostly retired group of scientists. The IPCC is highly criticized by supporters of the NIPCC for its ties to government and lack of research into causes of global warming and climate change due to natural variations. I introduce these two groups and their relationship to one another because any serious discussion of the environment must include credible dissenting views.

Key points of the IPCC Assessments

Between 1992 and 2014 there have been 5 IPCC assessments. Each assessment consists of reports from three working groups: Working Group I, The Science of Climate Change; Working Group II, Impacts, Adaptations and Mitigation of Climate Change: Scientific-Technical Analyses; and Working Group III, Economic and Social Dimensions of Climate Change. These voluminous reports increase in length and substance with each assessment report as more scientists and reviewers are added. For purposes of this book I will briefly summarize the contents of the first four of these assessment reports (AR1 through AR4) before moving on to the last report (AR5). These reports are readily accessible online in pdf format.[6]

1. Climate change now affects every part of the planet.
   a. More negative than positive impacts to crops
   b. Increasing degradation to coral reefs
   c. Higher risks posed by extreme events such as heat waves and coastal flooding
   d. Increased tree deaths in various regions
2. Climate change will increase the frequency and severity of extreme weather.
   a. Increased extreme precipitation events year-round in northern Europe
   b. Coastal flooding under conditions where sea level rises by 0.5 meters
   c. Increased heat waves especially in northern Africa

3. Meeting the scale of challenges requires adaptation and mitigation. Given historical and current [GHG] emissions, we're locked into additional warming. Adaptation is especially critical for coastal areas where those populations are most vulnerable. Adaptation alone cannot overcome all climate change impacts as at some point adaptation will become too costly and therefore will require significant mitigation efforts.
4. Rapid and steep reductions in GHG emissions can reduce the risks and costs, and an emphasis on timing matters. Limiting global temperature rise to 2 degrees C (the internationally agreed upon target) is achievable but only if countries act with urgency to significantly and rapidly reduce their emissions. The 2 degrees C temperature rise target would prevent many of the risks that become more severe the higher the temperature climbs.
    a. With each degree of warming renewable water resources will (predicted) decline 20 percent for each additional 7 percent of the global population.
    b. With warming greater than 2 degrees C there is a high risk of abrupt and irreversible changes to the structure of function of ecosystems and the oceans will become less efficient at absorbing CO_2 which would amplify the impact of warming.
    c. If global temperatures rise more than 4 degrees C—and there is a 20 percent chance of exceeding this level by 2100 even if existing emissions reduction commitments are implemented, there is a high confidence that we could witness adverse impacts

on agricultural production worldwide, extensive impacts to ecosystems, and increased risks of species extinction.

The fifth and last assessment report expands on previous assessments with new evidence of observed climate change. AR5 makes it very clear that human influence has been the dominant cause of the observed warming since the mid-twentieth century. Here is a summary of key points from this last IPCC assessment:[7]

- Each of the last three decades has been successively warmer at the Earth's surface than any preceding decade since 1850.
- As warming continues, the Oceans store the increase in energy in the climate system accounting for more than 90 percent of the energy accumulated between 1971 and 2010. It is virtually certain that the upper ocean (0-700 meters) warmed from 1971 to 2010.
- Over the last two decades, the Greenland and Antarctic ice sheets have been losing mass, and glaciers have continued to shrink almost worldwide. Arctic sea ice and Northern Hemisphere spring snow cover have continued to decrease in extent (high confidence).
- The rate of sea level rise since the mid-nineteenth century has been larger than the mean rate during the previous two millennia (high confidence). Between 1901 and 2010 global mean sea level rose by 0.19 meters (7½ inches)

- Atmospheric concentrations CO_2 have increased to unprecedented levels in the last 800,000 levels and have increased 40 percent since pre-industrial times primarily from fossil emissions and secondarily from net land use change emissions (deforestation and cultivation). The largest contribution to radiative forcing is caused by the increase in atmospheric concentration of CO_2 since 1750.
- The ocean has absorbed about 30 percent of the emitted anthropogenic CO_2 which has cause ocean acidification.
- Human influence on the climate system is evident from the increasing greenhouse gas concentrations in the atmosphere.

The message from the IPCC assessments is we must rapidly rein in GHG emissions now to prevent costly economic impacts in the future.

Key points of concern of the NIPCC in assessing the IPCC reports:[8]

- The IPCC is only trying to focus on the human cause and does not take seriously natural causes.
- Amidst all the research done by the IPCC, there is a lack of skepticism which they believe is at the core of scientific research. Where are the voices of dissent within the IPCC which are always a part of research findings?

- The NIPCC is concerned that many of those IPCC scientists are afraid to speak out for fear of losing their positions or grant funding and thus they wait until they are emeritus in order to speak out on the issues.
- The NIPCC is concerned that since the IPCC is government-sponsored, it is politically motivated and predisposed to believing that climate change is a problem in need of governmental and U.N. solutions.
- Whereas the reports of the United Nations' Intergovernmental Panel on Climate Change (IPCC) warn of a dangerous human effect on climate, NIPCC concludes the human effect is likely to be small relative to natural variability, and whatever small warming is likely to occur will produce benefits as well as costs.

Further findings:

- That warming from greenhouse gases will be so small as to be indiscernible from natural variability.
- The impact of modestly rising CO_2 levels on plants, animals and humans has been mostly positive.
- The costs of trying to limit emissions vastly exceed the benefits.

Specific reports the scientists found:[9]

- There is no scientific consensus on the human role in climate change.

- Future warming due to human greenhouse gases will likely be much less than IPCC forecasts.
- Carbon dioxide has not caused weather to become more extreme, polar ice and sea ice to melt, or sea level rise to accelerate. These were all false alarms.

For Public Policy, the NIPCCs report means:[10]

- Global warming is not a crisis. The threat was exaggerated.
- There is no need to reduce carbon dioxide emissions and no point in attempting to do so.
- It's time to repeal unnecessary and expensive policies.
- Future policies should aim at fostering economic growth to adapt to natural climate change.

What about those who still say global warming is a crisis?[11]

- The UN's new report walks back nearly a dozen earlier claims, contains more than a dozen errors, and tries to cover up new discoveries that contradict its earlier claims.
- The Environmental Protection Agency (EPA) relies heavily on the UN's reports for its finding that carbon dioxide is a pollutant. That finding is now falsified.
- Environmental groups refuse to admit they were wrong. It was never about the science for them.

The bottom line finding of the NIPCC is that global warming is an entirely manageable, if not beneficial, change in the climate.[12]

A Review of the Two Positions

IPCC	NIPCC
1. Focus is on anthropogenic causes of global warming.	1. Focus is on both the anthropogenic and natural climate cycle causes of global warming.
2. Warming from anthropogenic GHG emissions potentially a catastrophic crisis.	2. The warming from anthropogenic GHG emissions will be so small as to be indiscernible from natural variability. Global warming is not a crisis
3. There is a sense of urgency to reduce human GHG emissions.	3. There is no way to quantify the contribution of human activity to global warming so as to make meaningful global policy change.
4. There is a scientific consensus in the scientific literature that the human role is driving global warming.	4. There is no scientific consensus on the human role in global warming.

So Why do Scientists Disagree about Global Warming

For years, the literature on global warming was relatively one-sided, favoring the view that global warming was caused by human actions, specifically, increased emissions of CO_2. Increasingly, more scientists and journalists have become skeptical, partly because global warming, despite NOAA's claim to the contrary, has slowed and seemingly stopped; a claim made by investigations and studies, and in testimony before numerous congressional hearings. In this graph this pause in the warming of the latter part of the twentieth century can be seen from the **RSS** satellite data of atmospheric temperatures.

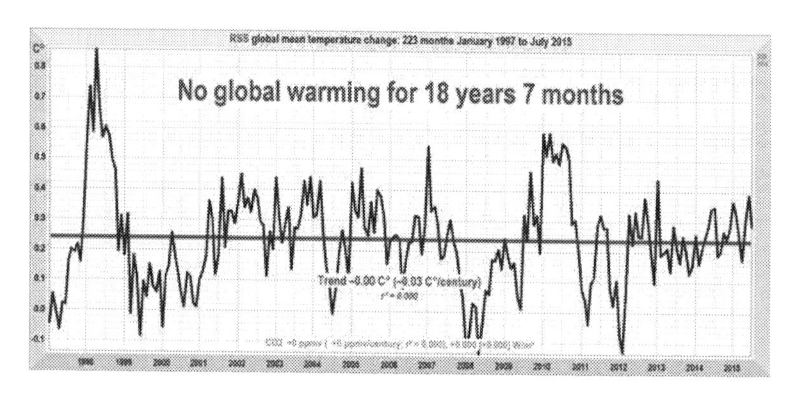

Graph Courtesy of Climate Depot

Another reason is that while global temperatures have remained constant so far in the first two decades of the twenty-first century, CO_2 concentrations as recorded by **Cape Grim** [13] has continued to rise and has surpassed the 400ppm level, considered by many just 30 years ago as the **tipping**

point concentration from which there could be no return from the increase in global warming.[14] At the beginning of the Industrial Revolution in the eighteenth century, CO_2 concentration was estimated at 275ppm. In 200 years CO_2 concentrations have increased by about 45 percent while temperatures have increased over that same period of time by about 1.6 degrees C. The following two graphs, courtesy of Cape Grim and the IPCC, illustrate these increases.

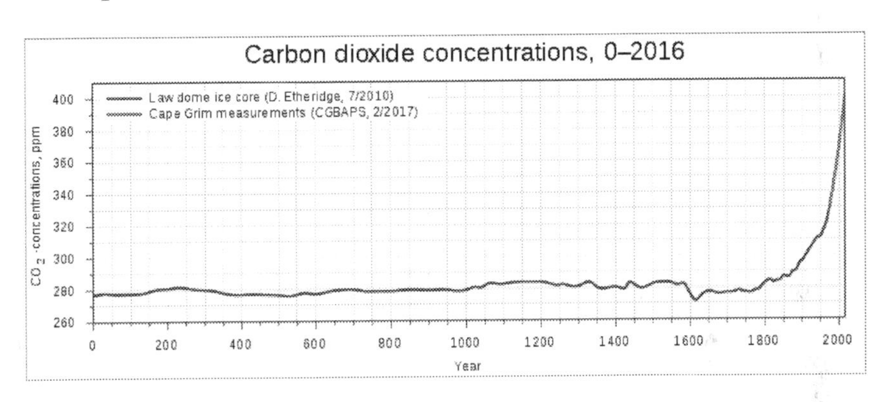

This increased concentration has not led to the warming predicted by modeling especially in the last forty years, as suggested by Dr. Roy Spencer using satellite data (RSS) and the University of Alabama, Huntsville (UAH).[15]

It is important to note here that global warming is measured most accurately in atmospheric temperatures, not surface temperatures that can be greatly influenced by **heat islands**, (urban development—buildings, concrete structures, parking lots, etc).

Scientists disagree on the science of global warming in large part based on a NASA claim from multiple studies in peer-reviewed scientific journals studies claiming that there is a 97 percent consensus among actively publishing scientists who agree that modern day global warming is most likely caused by human activity. These claims are supported by NASA based on four sources they cite:[16]

- J. Cook, et al, "Consensus on consensus: a synthesis of consensus estimates on human-caused global warming," *Environmental Research Letters*Vol. 11 No. 4, (13 April 2016); DOI:10.1088/1748-9326/11/4/048002
- W. R. L. Anderegg, "Expert Credibility in Climate Change," *Proceedings of the National Academy of Sciences* Vol. 107 No. 27, 12107-12109 (21 June 2010); DOI: 10.1073/pnas.1003187107.
- P. T. Doran & M. K. Zimmerman, "Examining the Scientific Consensus on Climate Change," *Eos Transactions American Geophysical Union* Vol. 90 Issue 3 (2009), 22; DOI: 10.1029/2009EO030002.
- N. Oreskes, "Beyond the Ivory Tower: The Scientific Consensus on Climate Change," *Science* Vol. 306 no. 5702, p. 1686 (3 December 2004); DOI: 10.1126/science.1103618.

The following are excerpts I have taken from a Heartland Institute Presentation to a July 2016 meeting of Doctors for Disaster Preparedness that show many scientists such as Judith Curry (former chair of Earth and Atmospheric Sciences, University of Victoria), and Roy Spencer (a meteorologist and and principle research scientist, University of Alabama, Huntsville) among many others; disagree with the conclusions of these four sources.[17] We will look at them in reverse order so they will be in chronological order.

The first one is Naomi Oreskes, a historian at the University of California, San Diego, who claims to have examined 928 abstracts from papers published from 1993 to 2003 and said that 75 percent of the abstracts either implicitly or explicitly supported the IPCCs view that humans are responsible for most of the observed warming over the previous fifty years. The problem with this source is that it is not peer-reviewed. Though NASA said it appeared in a peer-reviewed publication, which it did, but it is an op-ed piece in science, an essay, an opinion piece that itself was never peer-reviewed. In the abstracts studied she searched for global climate change instead of climate change. By doing that she left out 92 percent of all the articles that talk about climate change including many that talk about natural causes of climate change. The claim by some scientists is that because she is not a scientist, she misinterpreted many of those abstracts were meant to say. Analysis of an abstract of an article is not considered a valid research method for content as abstracts are jammed full of key words that will attract search attention from other academics and is meant to "sell" citing that article in other's to gain tenure and ranking. She never stated how many of the articles agreed with the IPCCs definition of climate change, which is a very specific definition.

The second study on consensus cited by NASA by Maggie Kendall Zimmerman and Peter Doran was a thesis paper written by Maggie. Peter Doran was her thesis advisor. Putting his name on it got it published in a peer-reviewed journal in 2009. Her study consisted of a two minute online survey of earth scientists working for government agencies and universities. Many scientists find fault with this study as it asked the wrong people, excluding solar scientists, space scientists, physicists, meteorologists, and astronomers. Only 5 percent of respondents self-identified as climate scientists.

The third study is a by another college student, William Anderegg in 2010 doing a paper for his college professor who puts his name on it and gets it published in the Proceedings of the National Academy of Sciences. Anderegg's study found "97% of climate researchers most actively publishing in the field" support the UNs IPCC findings. He looked at the fifty most prolific alarmists scientists who were published an average of 408 times each and found that 98 percent of those fifty scientists agreed with the IPCC. Many scientists make the point that being published less often does not make those scientists less credible and being published 408 times is not necessarily evidence of quality but possibly evidence of fraud in a broken peer-reviewed process as more and more published articles often list twenty or more authors. As those articles get published, those twenty or more authors get to add another to their resume. Clearly the peer-review process that includes that many endorsing scientists is suspect of being more of a friend review. And obviously 98 percent of fifty scientists is no consensus.

The fourth study cited by NASA is from an Australian blogger, John Cook, who published in Environmental Research Letters and claimed to find 97 percent of articles published from 1991 to 2011 endorsed the view that human activity is responsible for some warming. This is another abstract counting exercise subject to the same errors in data gathering and analysis of the content of those scientific papers as Naomi Oreske's study. When others like economist

David Legates, solar scientist Willie Soon, and journalist Christopher Monckton looked at the same data Cook used; they concluded that just 0.3 percent of the scientists in Cook's study endorsed the "consensus" that most warming since 1950 is anthropogenic.

While there is no clear scientific evidence that a consensus exists, there is, however, evidence of disagreement among scientists as to a consensus.[18] In a 2012 survey, consisting of 54 scientific questions, of international climate scientists by Dennis Bray and Hans von Storch it was found that about a third of them identified as alarmists, a third skeptics, and a third said they were uncertain. In 2015, Kenneth Richard, Associate Professor of Public and Environmental Affairs, Indiana University; presented 250 peer-reviewed papers finding natural factors a dominant climate driver. An interesting study, in my mind, would be to study if there is any kind of consensus among scientists on natural factors driving climate change.

The best evidence of disagreement among scientists is Art Robinson's (an American biochemist) Oregon Petition signed by 31,478 American scientists including 9021 with Ph.D.s saying in part "There is no convincing scientific evidence that human release of $CO_{2,}$ methane, or other greenhouse gases is causing or will, in the foreseeable future, cause catastrophic heating of the Earth's atmosphere and disruption of the Earth's climate."[19] Those who find fault with this petition say "it was created by group of individuals with political motivations presented with almost no accountability regarding the authenticity of its signatures, and asks only that you have received an undergraduate degree in any science to sign." (Kasprak 2016) As I have previously cautioned, we must pay attention to how the "question or statement is made." Those critical of the Oregon Petition claim those 31,478 American scientists "reject the concept of anthropogenic global warming," however the claim made by those signing the petition is *"There is **no convincing scientific evidence** that human release of CO_2, methane, or other greenhouse gases is causing or will, in the foreseeable future, **cause catastrophic heating** of the Earth's atmosphere and **disruption of the Earth's climate.**"*[20] This petition is not a rejection of the **concept** anthropogenic global warming, rather a judgment of the degree of its impact on global warming.

Conclusion

Why do scientists disagree? Here are four reasons:

1. Climate science is a complex interdisciplinary science that requires insights from many fields of scientific expertise that have not been mastered by any of them. These disciplines include:
    - Archaeology
    - Geology
    - Physics
    - Statistics
    - Science of forecasting
    - Chemistry

    Because virtually no one in the field of "Climate Science" admit to expertise in the broad range of these scientific disciplines, disagreements should be expected as different conclusions and insights abound.

2. There is insufficient observational evidence, and therefore there exists fundamental doubts on how to interpret data and design models to predict the future climate. In years to come we will still be debating what the human impact will be on climate.
3. The IPCC which was created to find the human impact on global climate, and not include the impact of natural variation impact is not considered a credible source among many scientists. Those who find fault with the IPCC claim it was created to give the appearance of consensus which turned it into an advocacy agency so the UN could do something about the problem. Rajendra Pachauri, former head of the IPCC admitted that UN climate reports were "tailored to meet the needs of governments."[21] Because the process of these reports is not open to debate, or to opposing points of view, many IPCC scientists like *Dr. Kenneth Green a Working Group 1 expert reviewer for the United Nations';* South African Nuclear Physicist and Chemical Engineer, Dr. Philip Lloyd, a UN IPCC co-coordinating lead author who has authored over 150 refereed publications; *Alabama State Climatologist Dr. John Christy of the University of Alabama in Huntsville, who served as a UN IPCC lead author in 2001 for the 3rd assessment report and detailed how he personally witnessed UN scientists attempting to distort the science*

for political purposes. to name a few, have all resigned from the IPCC process.[22]

4. Climate scientists can be biased based on careerism, grant-seeking, and political views. The East Anglia email scandal (Climategate) which will be discussed in chapter 11, is a good example of this bias.

CHAPTER SIX

THE EPA

"If you follow your heart, if you listen to your gut, and if you extend your hand to help another, not for any agenda, but for the sake of humanity, you are going to find the truth."
—Erin Brockovich

Introduction

As the impact we humans were having on our environment became obvious, most notably through polluted skies and waterways, the United States government became involved by passing the National Environmental Policy Act of 1969. In order to consolidate many of the federal government's initial environmental responsibilities under one agency, President Richard Nixon signed an Executive Order on December 2, 1970 establishing the Environmental Protection Agency (EPA).

An Administrator appointed by the President and approved by Congress leads the EPA. Though this position is not officially a part of the President's Cabinet, the Administrator normally holds Cabinet rank and is charged with enforcing the laws of Congress under direct control of the President.[1]

According to its charter, the mission of the EPA is to protect both human health and the environment within the United States. Its immediate goals within this broad mission are to (1) regulate and enforce primarily through fines, those regulations passed by Congress to clean up our polluted air and waterways; (2) conduct environmental assessment and related research; and (3) educate the American people.[2] Thus the EPA is responsible for maintaining and enforcing national standards under a variety of environmental laws, in consolidation with state, tribal and local governments. Since the EPA falls under the executive branch of our government, it is important that Constitutional checks and balances be in place to counter a concentration of power within the EPA. This idea will be further explored later in this chapter.

My goal in writing this book was to keep politics from the environmental debate to the extent possible and this remains my aim in this chapter to the extent possible. However, any discussion of a government agency such as the EPA, presents a challenge to this goal due to its basis in political appointment. I will attempt to remain objective in my assessment of the EPA.

The EPAs History: A Series of Successes

Initially the EPA effectively tackled major environmental concerns such as water pollution in our lakes and rivers, hazardous auto emissions, and the use of the insecticide DDT. Americans have benefitted from these efforts and arguably the EPA has set an example at the international level on how to reverse the contamination we have inflicted upon our environment. Noteworthy EPA accomplishments include the following:

1. Banning DDT
2. Passing the Clean Air Act
3. Reducing the pollutants that cause smog and acid rain by supplying lead-free gasoline and catalytic converters in automobiles (According to the California EPA, new cars today produce 98 percent cleaner emissions than in 1970 in terms of smog-forming pollutants[3])
4. Reducing **acid rain** by cleaning coal-burning plants with industrial smoke stack scrubbers (these help remove sulfur that is emitted when coal is burned)
5. Establishing the Superfund Program, which cleans hazardous waste sites and sets standards for water quality and the reduction of mercury from power plant emissions?

In short, the EPAs work has resulted in cleaner air, cleaner drinking water, cleaner lakes and rivers, with hundreds of previously declared impaired bodies of water now meeting water quality standards in this country. Other countries have followed and in many cases have taken their own leadership roles internationally. Unfortunately, many others have not.

At Issue: Controversial Policies of the EPA

Though the EPA has made significant strides in cleaning up our air and waterways, there is growing concern that it has also begun operating outside the scope of its charter, even to the point that Congress has ceded direct oversight of a range of EPA issues. Any discussion of this organization would be incomplete without a reflection on this controversy, because we as active environmentalists must work within the EPAs mandates and bear the costs associated with their directives, both in dollars and quality of life. In chapter 11 I discuss these costs in further detail.

My aim is not to downplay or criticize the many highly skilled EPA employees who work with dedication and to improve our environment, but the leadership of the EPA has increasingly been the target of concern and investigation for perceived overreaches of authority.

In 1991 a panel of outside scientists was brought in to review accusations that the EPA often shields key research from peer review. As EPA Administrator William Reilly (1989-1993) acknowledged, "Scientific data have not always been featured prominently in environmental efforts and have sometimes been ignored even when available."[4] A review of some of these shielding issues and accusations follows.

The EPA and DDT

In 1972 the EPA banned the general use of DDT having determined through "inquiry" that the continued use of it posed unacceptable risks to the environment and potential harm to human health.[5] Extensive hearings on DDT occurred before an EPA administrative law judge in 1971. The EPA hearing examiner, Judge Edmund Sweeney, concluded that 1) "DDT is not a carcinogenic hazard to man, 2) DDT is not a **mutagenic** or **teratogenic** hazard to man, and 3) the use of DDT under the regulations involved here do not have a deleterious effect on freshwater fish, estuarine organisms, wild birds or other wildlife."[6] Disregarding his own examiner, EPA Director William Ruckelshaus banned DDT in 1972.

Among other concerns at the time, DDT was blamed for the decline in the bald eagle and peregrine falcon populations. Yet the bald eagles were reportedly threatened with extinction in 1921, twenty-five years before widespread use of DDT. After fifteen years of heavy and widespread usage of DDT, Audubon Society ornithologists counted 25 percent more eagles per observer in 1960 than during the pre-DDT 1941 bird census.[7] U.S. Fish and Wildlife Service biologists fed large doses of DDT to captive bald eagles for 112 days and concluded that "DDT residues encountered by eagles in the environment would not adversely affect eagles or their eggs."[8]

As for the decline in the peregrine falcon population, it occurred long before the DDT years. During the 1960's, peregrines in northern Canada were "reproducing normally," even though they contained 30 times more DDT, than the Midwestern peregrines that were allegedly eradicated by those chemicals.[9]

The point here is that DDT is not necessarily a harmful agent but that without congressional oversight one can conclude the EPAs inquiries are just inquiries rather than actual scientific studies and should not translate into policy directives.

The EPA and CO_2

When President Obama could not get his Cap and Trade Bill through Congress, he used an executive order (EO) giving the EPA the authority to declare CO_2 a pollutant. After this action, the EPA could use that order to regulate/tax CO_2 emissions.

Under the Clean Air Act the EPA, with the backing of President Obama's EO, justifies its attempts to control CO_2 emissions because as a GHG and its suspected role in global warming, it should be considered a "pollutant" to clean air. It is arguably a stretch to label CO_2, a necessary nutrient to all plant life, a pollutant. It is important to remember that CO_2 is a colorless, odorless gas whose role in global warming is well known but the impact of its anthropogenic involvement has not been scientifically quantified (in a paper published by EPA scientist Fred Haynie).[10] Allowing the EPA to regulate CO_2 as part of the Clean Air Act (without Congressional oversight), has far reaching political ramifications; both environmentally and for sure economically.

The appointment of Scott Pruitt by President Donald Trump as the administrator of the EPA has been controversial because Mr. Pruitt has been quoted as stating he does not believe that CO_2 is a primary contributor to global warming and was reported by CNBC as saying: "I think that measuring with precision human activity on the climate is something very challenging to do and there's tremendous disagreement about the degree of impact, so no, I would not agree that it's a primary contributor to the global warming. We need to continue the debate and continue the review and the analysis." (DiChristopher 2017) Scott Pruitt has since been replaced at the time of this writing by Andrew Wheeler.

Pruitt's statement contradicted the public position of the EPA agency he then lead. The EPA's webpage on the causes of climate change states "Carbon dioxide is the primary greenhouse gas that is contributing to recent climate change." This view is also held by the IPCC as presented in the last chapter. Based on the scientific evidence presented in chapter 6, there appears to be considerable uncertainty in the contribution of CO_2 to global warming.

The EPA and Coal

Coal remains the major source of power generation in this country and throughout the world. If not for coal, the Industrial Age certainly would have been greatly postponed and would have progressed differently. It was a cheap and abundant fuel that drove rapid increases in the generation of light, heat, and transportation which vastly improved and continues to improve the quality of our lives. To this day it remains a cheap and abundant fuel source around the world. But coal-reliance as an energy policy has a big down side that must be addressed if we are concerned about both CO_2 and air contaminant emissions as coal must be burned to derive the energy it contains. Unless clean-burning procedures are incorporated, this burning process pollutes the air with nitrogen oxides and sulfur dioxide, particulate matter (smog), mercury, and dozens of other substances know hazardous to human health. Because the combustion of coal releases CO_2 and the pollutants just mentioned into the atmosphere, the mining and burning of coal is justifiably a serious concern for the EPA.

In consideration of our society's other broad interests, the coal industry is also a major component of our economy and energy policy and has been for a hundred years. Should the EPA, lead by a political appointee and created under a Presidential EO, have the power to control this industry, and the power to potentially shut it down, affecting the economy and lives of so many citizens? Therein lies the controversy.

There are some compromise solutions to this controversy, in that clean burning technologies are available, being used, and continue to be developed to deal with this downside of the industry. These include **gasification**, scrubbing flue gases to remove pollutants, and **carbon capture and storage** (discussed in chapter 10), to name a few. The successful cleanup of auto emissions is an EPA success story, and that story can be told for the coal burning industry as well. Coal can be burned cleanly and by its own admission the EPA demonstrates that these technologies have made today's coal-based generating plants 77 percent cleaner on the basis of regulated emissions per unit of energy produced.[11] Still (good or bad) the EPA continues to push industry-crippling regulations, primarily on CO_2 emissions reductions on the coal industry.

Interestingly, the EPA faces increasing opposition from a source that has so far protected the coal industry, at least to a limited degree. Harvard Constitutional Law Professor Laurence Tribe writes:[12] *"The [EPA's rule] demonstrates the risk of allowing an unaccountable administrative agency to 'make' law and attempt to impose the burden of global climate change on an unlucky and unfortunate few.... The EPA's singling out of a mere handful of emitters and limiting (or curtailing) their property is exactly the type of overreaching the Fifth Amendment seeks to prevent."* Professor Tribe's argument that EPA rulings are unconstitutional is one of the strongest charges made against the agency's Clean Power Plan. However, it is important to note that Professor Tribe is on a retainer for the coal company Peabody Energy.

The thought that "the risk of allowing an unaccountable administrative agency to 'make' law and attempt to impose the burden of global climate change on an unlucky and unfortunate few," is an interesting one that needs to be considered on an international scale as well.

The Constitutionality of EPA Actions in Question

The Controversies

It has been said by many, including some in Congress, that the EPA has circumvented the Constitution given that their edicts or rulings are not representative of the people, because these rulings have not been subject to Congressional oversight and therefore violate the separation of powers required by our Constitution. This circumvention makes the EPA, in and of itself; the executive, legislative, and judiciary all wrapped into one.

Also, there is the question of whether some of the actions of the EPA violate the Administrative Procedures Act, which governs the writing of regulations and takes a dim view of outside, special-interest organizations secretly drafting government rules. Freedom of Information Act requests have led to documents that detail evidence of collusion between the EPA and green activist groups such as the National Resources Defense Council (NRDC), and the Sierra Club. "These documents further disclosed that EPA Administrator, Lisa Jackson, used a.k.a. 'Richard Windsor' as her private Verizon account to e-mail directly, off the record, green lobbyists such as Sierra Club's Michael Brune."[13] The Federal Advisory Committee Act also requires federal officials to interact in a prescribed and open manner with private entities such as lobbyists and environmental groups.

Considering these controversies, a balanced review of charges against the EPA made by those adversely affected by its regulations is appropriate.

In November 2013, in one such charge, the Congressional Committee on Energy and Commerce sent a letter to the EPA Administrator, Gina McCarthy, expressing its concern that the agency was ignoring reports from its own scientists who reviewed carbon emissions limits for new power plants. The Congressional Committee claimed the agency rushed through the regulatory process and the underlying science of the ruling lacked adequate peer review. "We are concerned about the agency's apparent disregard for the concerns of its science advisors," the lawmakers wrote. "Science is a valuable tool to help policymakers navigate complex issues."[14] Their argument was, of course, that the EPA was disregarding or ignoring "inconvenient" facts in order to prevent frank discussion and debate.[15]

Upon closer examination of the Committee's letter it is apparent that legality is at the bottom of the Committee's concern. Current rules require the EPA to regulate with guidelines of "currently available commercial technologies."[16] The EPA's proposed power plant emissions limits, which are based on the Clean Air Act, would require that all new coal-fired power plants be built using **Carbon Capture and Storage** (CCS) technology. However, CCS is not yet a commercially proven or commercially viable technology. Thus the committee was questioning whether the EPAs standards were beyond the scope of its legal authority.

In another claim that the EPA was exceeding its authority, the Utility Air Regulation Group filed charges against the EPA in the Supreme Court of the United States in December 2013. The Regulatory Group claimed that the EPA was exceeding its authority by requiring those permits required under the Clean Air Act for new motor vehicles also were being extended to stationary sources (such as boilers, heaters, furnaces, and any other equipment that combusts carbon bearing fuels) that emit greenhouse gases.[17] The rules put forward by the EPA in this regard would fuel the argument that such EPA action would require a rewrite of the statute which it should not be entitled to.

At the time of this writing, the city of Flint, Michigan is in crisis over its lead-contaminated water supply. All indications are that both the federal and state EPAs as well as local government knew there was a problem from almost three years of water quality complaints from Flint's residents but did little to nothing to prevent. A few years ago the EPA admitted poisoning the Animas River in Colorado due to its own negligence. To date no agency or individual has been held accountable for these two man-made environmental tragedies resulting from human failure. Justice in these investigations will be slow in coming, if it ever does. Hopefully the EPA is not more interested in controlling industry, than in its own malfeasance. We would all be better served by an EPA that is more concerned with the immediate health concerns of the water we drink and the air we breathe than with its prioritization with CO_2 emissions, which by the way, this country has already reduced greatly with its conversion in much of its energy production to natural gas.[18]

Conclusion

The EPA has been a very successful government agency in cleaning up air and water in the United States. However, it is an agency that has grown in size and scope beyond its original charter. As I presented at the beginning of this chapter, the EPA falls under the executive branch of our government, and therefore it is important that checks and balances be in place to counter a concentration of power within the EPA. In consideration of the abovementioned cases as well as other examples not specifically mentioned, the legislative branch of our government has not effectively checked the EPAs increasing growth and scope of authority. Without open debate encouraged by all sides of these issues, the EPA and its supporters can project a program that may not be in the best interest of the American people who will ultimately bear the costs, burdens and sacrifices of EPA policy directives. There is much at stake in this matter as the EPA, through its actions, can have a significant impact on our industry, economy, and our quality of life.

My purpose in airing some of the EPAs "dirty laundry" is not to take a side, but to expose a side of the EPA not generally known or reported. The EPA has done great things for the environment, yet over-reach and abuse of power are problems that must be confronted. The resolution of this controversy is political and therefore beyond the scope of this book. The point I make with this chapter is "reader beware."

CHAPTER SEVEN

CONSERVATION ISSUES OF CAP AND TRADE, RECYCLING, AND REUSING

"Nobody in this country realizes that cap and trade is a tax, and it's a great big one."
—Rep. John Dingell (D)

"There is no such thing as 'away.' When we throw anything away it must go somewhere."
—Annie Leonard, Executive Director of Greenpeace USA, and proponent of sustainability

Introduction

In his 2013 State of the Union Speech, President Obama "warned that Americans must take steps now to cut our emissions of the dangerous carbon pollution that threatens our planet," appealed to the "overwhelming judgment of science" and referred to "recent weather extremes in the

United States to include; Superstorm Sandy, the most severe drought in decades, and the worst wildfires some states have ever seen…"[1] As evidence for man-made global warming, the President said, "The good news is we can make meaningful progress on this issue while driving strong economic growth." To that inference, and in a nod towards climate change bipartisanship, the president mentioned the collaborative efforts of Senators John McCain and Joe Lieberman in their proposed Climate Stewardship Act of 2003[2] to develop a way to promote a market-based solution to climate change. That market-based solution is the basis for the concept of **Cap and Trade**.

Until that time the only previous approach to dealing with climate change was the Kyoto Protocol negotiated by President Clinton's administration in 1997 which called for a voluntary reduction of emissions to 1990 levels. Though ratified by 100 countries, President Clinton, who believed that a reduction of greenhouse gases was the wise approach to dealing with climate change issues, signed it but could not convince the U.S. Congress to ratify it. Congress cited "possible damage to the U.S. economy required by compliance with it."[3] President George W. Bush rejected the Kyoto Protocol declaring it as unfounded in science and too expensive to pursue.[4] The 2003 McCain/Lieberman Climate Stewardship Act previously mentioned was in contrast to the voluntary measures of Kyoto and required domestic mandatory and economy-wide emission reductions. The bipartisan nature of this bill which featured a climate research program, an emissions registry, and a trading program based on government distributed emission "allowances" for each sector based on the emissions it had for the base year (2000).[5] In subsequent years, the limit would be reduced to the 1990 emission levels. The McCain-Lieberman Climate Stewardship Act ultimately did not pass, however, it set in motion an awareness on limiting GHG emissions and represented one of the most significant domestic proposals for a climate change policy up to that time. This led to a limited version of cap and trade now in use in California. Europe and Japan have also implemented limited forms of cap and trade.

The idea of requiring an "allowance" for emission of CO_2 and other greenhouse gases would be based on the unit measure of "a ton" of carbon. Why measure a gas in "tons" rather than some other measure is a logical question to ask? (No one weighs the gas that comes out of a chimney or exhaust pipe.) The short answer is a "ton of carbon" means enough CO_2 molecules so that the combined mass of all the carbon (a solid like a chunk of coal) in those molecules equals one ton. When the carbon atom burns it picks up two atoms of oxygen, so carbon is roughly a third of the weight of a CO_2 molecule. When cap and trade allowances are made producers or importers of industrial greenhouse gases would be required to obtain an allowance for each amount equivalent to a ton of carbon dioxide that would be produced or imported (rather than emitted). That was the plan at that time, year 2003.

So, how much does a ton of carbon cost?

The cost of carbon is nothing more than a tax, the idea of which is to put a simple price on emissions of carbon dioxide or other greenhouse gases for some dollar amount per ton of that particular GHG (for simplification we'll just use CO_2) be determined at some time based on clean up costs to the environment and the market value of a ton of CO_2 when it traded from an emitter who doesn't need it to another who does. These "costs" will be very arbitrary at the beginning since actual costs will not be well known and can only be estimated. The economic argument favors pricing because of its efficiency. When CO_2 emissions are taxed, the alternative energy sources are helped equally as the cost to the carbon emitting industries rise through cap and trade taxation. This allows the alternative renewable and carbon free energies, whose initial costs are high, a more favorable market in which to compete for energy dollars. Under the current taxation system the renewable energy industries enjoy a more competitive environment due to taxpayer subsidies. These alternative renewable and carbon free energies would include, wind, solar, hydroelectric, home insulation, and higher mileage automobiles to name a few.

The question remains, how much does a ton of carbon cost? Data used to determine the cost of CO_2 emissions is very limited, but as an example, we can use Tokyo, Japan where cap and trade programs have been implemented in a program in that city. Carbon credits sold for $142 per ton of CO_2 emissions in initial trading in 2010.[6]

The process surrounding assigning these allocations and credits toward future use is complicated and beyond the scope of this book. The point made is an overall emission rate would be the goal for all in establishing a baseline from which future year emissions could be reduced through reduced allocations. A market would then naturally follow where Company A could sell unused allowances to Company B who was having difficulty staying within its allocated allowance. The net result, in a perfect process, would then be emissions that were quantifiable, manageable, and reducible. One can only imagine the government bureaucracy this concept of cap and trade would create; which most probably would have unforeseen unintended consequences that typically follow bureaucracies. One example that would necessarily follow is "how does the government handle violators" or "fraudulent use" of the allocations. The workings of government to develop the language required to get a bill into law, are not for us ordinary folk to comprehend. The main point to remember is that we who benefit from this plan to control and manage CO_2 emissions will pay this tax, based on our own lifestyle needs, which in turn will be an incentive to reduce our consumption of the products which produce these emissions. From heavy industry to the individual, all would be "required" to work within an energy allowance (that would be controlled by the government). Another unintended consequence arises: do individuals who own larger homes get a higher energy allowance?

Globally, a major concern that continues is the ability and/or willingness for those of us in the developed world to pay this consumption tax to cover for third world countries whose economies are unable to buy allocations until they can afford the cleaner energy technologies. The counter viewpoint of the more extreme elements of the environmental movement is that since the developed world has benefitted from the industries that have created those carbon emissions, it would only be fair to subsidize those developing countries until they can "catch up" in producing their own more efficient energy technology. This is a legitimate argument, and would require much economic sacrifice in the developed world to subsidize this disparity. And what would be the time frame for these expected sacrifices made to effect such change? Further it would take a "World Government" to implement an international cap and trade program. That in itself is a seemingly impossible transition in the near term. A compromise in my mind would be for the developed world to offer and be offered incentives to invest in green energies in the developing world. That to me seems much more feasible as many of these developing countries such as those in Africa are rich with resources that could be used to develop the cleaner energy infra structures. Clearly it would require the participation of dedicated leadership of those countries in the developing Third World. Again politics will get in the way of this process, though politics must be part of the solution.

In the U.S., as mentioned previously in this chapter, California has implemented a limited cap and trade program. This program began in 2013 as part of the states major policy changes to lower its GHG emissions. Their cap and trade program only applies to "large" (25,000 tons of CO_2 or more per year) power plants, industrial plants, and fuel distributors (e.g. natural gas and petroleum). This includes about 450 businesses for which the program is mandatory. California currently taxes CO_2 at $15 per metric ton (May 2018). To give an example, if a business estimates its cost for pollution control equipment to be lower than $15 per reduced ton, it would buy and install the equipment. If the cost is higher, it would buy the needed allowances. Businesses that cut their pollution can sell allowances to businesses that pollute more, or bank them for future use. California's cap and trade program has been claimed a success by law makers both in terms of increased state tax revenues and CO_2 emission reductions to 1990 levels two years earlier than their 2020 goal.

Of course, there are costs to all who live in California, because those 450 businesses will pay no tax as that tax is passed on to the consumer. Unfortunately it is the lower income population that pays proportionally greater amounts of their income for that carbon tax. For example, in December 2018, an average cost of a regular gallon of gas in California was $3.55 and a gallon of regular of gas in Missouri was $1.83. (Of course that price difference is not all tax as California's prices also contain additional costs for seasonal additives.)

Other countries or regions that have already passed "limited" (limits on only a portion of their carbon emissions) cap and trade laws include Australia, New Zealand, South Korea, the European Union, and the province of Quebec in Canada.

Recycling and Reusing The Waste From Human Activity

In 2012 the citizens of this planet produced about 2.6 Trillion pounds of trash[7] and this amount is constantly increasing. Where does it all go? Much of it goes into landfills, much litters our cities, rivers and lakes while far too little is **composted**, **recycled** or **reused** which would in most cases save money, energy, and resources. Almost half of this trash comes from organic waste, most of which ends up in landfills, which can be considered composting to some extent, since the organic portion of the trash land filled will return nutrients back into the soil. Important to this environmental topic and goal of sustaining a clean environment is how to deal with all this trash and garbage.

What is the composition of human waste?[8]

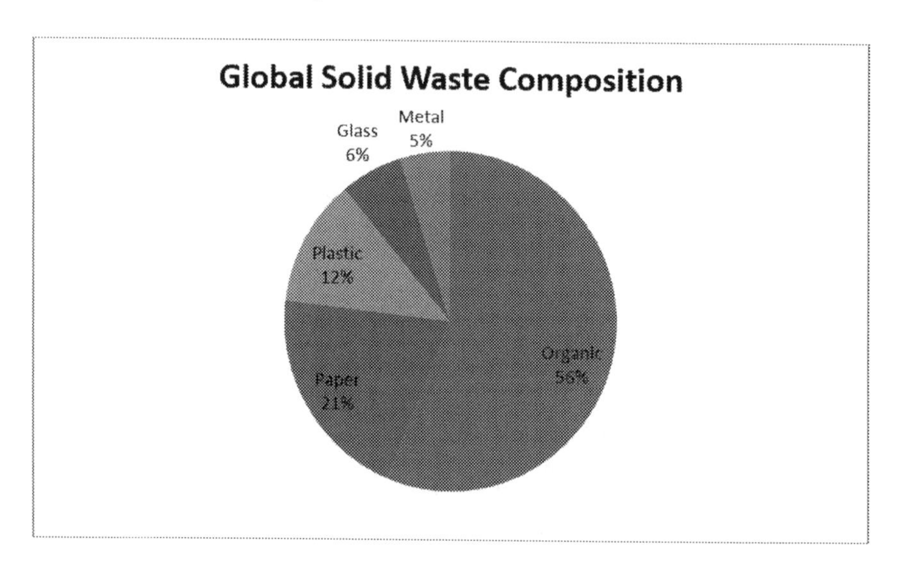

Organic waste includes material that is **biodegradable**, which means waste that comes from either a plant or animal. This would primarily be made up of vegetable and fruit debris, bones, and human waste which are generally quickly broken down by other organisms into nutrient material and returned to the soil, lakes, rivers, streams, and oceans. Not included in this composition of human waste is the hazardous waste humans produce from nuclear and chemical plants. "Other" waste includes textiles, leather, **e-waste** and other **inert** waste.

Which countries produce the most waste?[9]

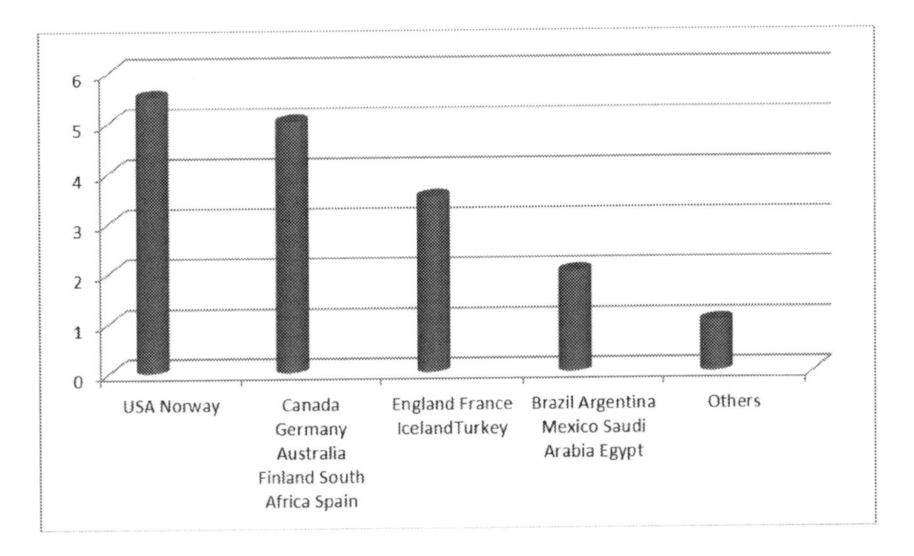

Quite clearly and expectedly so, developed countries produce the most waste, but then also have the highest collection and **R3** (recycle, reuse, and reduce consumption,) rates.

Is CO_2 a Human Waste?

I will save a more detailed description of CO_2 as a human waste for a later chapter, but it warrants a short discussion at this point. We have good evidence that CO_2 is a major greenhouse gas that warms the atmosphere of the planet. We know from chapter 3 that our climate is sensitive, albeit to an unknown degree, to CO_2 concentrations, but due to climate forcing factors that were presented in chapter 4; we are not sure to what degree. We know that these concentrations have increased from pre-industrial levels which averaged an estimated (primarily from ice core data) at 275ppm (parts per million) to about 396ppm in 2013, and because of the natural balance of CO_2 emission and absorption exhibited during recent past centuries, much of this increase is probably, but to an unknown degree, due to anthropogenic or human emissions primarily through the burning of fossil fuels. In the last 20 years CO_2 emissions have increased about 20 percent to this nearly 400ppm with no appreciable global warming.[10]

Though we hear claims of "this is the hottest year on record," those claims are based on very small increments that generally lie within the **margin of error** of the data presented. It may be of interest to point out in comparison to the atmospheric concentration of CO_2 at close to 400ppm, each exhalation of human breath has a concentration of about 40,000ppm of CO_2.[11] In dealing with CO_2 as a human waste we must be careful not to malign all the great benefits provided by this gas. All life on Earth depends on it and therefore this gas is a nutrient, at least up to certain unknown levels.

Recycling

Recycling is an old concept practiced for economic reasons throughout human history, but more recently practiced for environmental reasons. For many, recycling is habit that has become an "unqualified good." However, recycling is in some cases not an unqualified good and not an environmentally sound practice as other possible considerations. Recycling is basically separating our waste by category to be converted (through various forms of treatment processes) into reusable material. Though it sounds like an environmentally responsible activity, it has its own environmental drawbacks, since this effort takes additional consumable energy in its collection, transportation, and reprocessing procedures. For example, a plastic bottle in the recycle bin has a long journey ahead of it. First it goes to a collection facility to be inspected for contaminants; then it is washed and ground into flakes which are dried and melted into plastic lava, and then filtered for impurities. Finally, the lava strands are cooled in water and chopped into pellets that can go to market. Often times this "separated" waste or trash is not in fact recycled at all but landfilled when the market-driven costs of landfilling are cheaper than the cost of recycling. Though actual data on "recycled" material that ends up in landfills is not known, a December 2018 *National Geographic* article claimed only 9 percent of plastic is recycled.[12]

Transportation costs to distant processing facilities become significant as well as the costs associated with processing. Of course there is an environmental cost of using landfills that require large amounts of land usually distant from metropolitan areas and do "pollute" in their own right. It is difficult, if not impossible, to compute this cost of landfilling which therefore tends to make this cost economically easier to be ignored.

Therefore market forces play a big role in what, in reality, becomes recycled. For example, an ABC news report stated that glass, metal, and plastic recycling costs New York City $240 per ton, almost double what it costs to just throw it away[13]. With most city governments facing budget constraints, there is increased pressure to reduce or greatly curtail recycling programs especially when market prices on these recycle commodities are low. As the price rises it becomes more economical then to "recycle." However, "recycling" has become an environmental mindset to some, who very much resist or just ignore the cost savings of "just throwing it away" based on the belief that those environmental landfill costs, such as land required, and the supporting infrastructure of landfilling, are a significant consideration. The resistance to "just throw it away" often does not consider the journey a recycled item must take to get back into a usable form for human consumption or need. The energy costs of transportation alone in the recycling process can easily exceed those same costs of just "take it to the dump."

The question asked quite often among those debating or considering recycling plastic bottles is, does it cost more to recycle a plastic bottle or to make a new one. In the United States alone, about 29 billion plastic bottles are produced each year which requires the equivalent of 17 million barrels of crude oil for that process. In an effort to rate plastic, not only as an alternative packaging form, but also as a safe form of packaging (that would lend itself more uniform for recycling) a plastic classification called **Polyethylene Terephthalate** or **PET** was produced. Though recycling costs of PET are competitive with aluminum and glass, of the 2.7 million tons of plastic PET bottles on U.S. shelves in 2006, 80 percent still went to landfills.[14] Sadly too many plastic containers that are "just thrown away" end up in our waterways and oceans, and that is certainly not what is meant by "landfilling."

As mentioned, and setting aside the environmental concerns, the economics of recycling depends on the costs of raw materials and the cost of recycling versus the cost of disposal; these latter costs depend on cost fluctuations based on different metropolitan areas in proximity to recycling centers and the price to unload in local landfills. A University of California, Berkeley study "estimated that areas like Los Angeles and San Francisco could gain an economic benefit of $200 a ton for recycling instead of dumping."[15] Nonetheless, the cost of recycling a bottle versus making a new one simply varies, depending where the bottle is and what the volatile price of oil happens to be.

Reusing

Reusing is most probably the oldest and most environmentally friendly way to conserve costs, energy, and landfilling. I am reminded from the days of my youth, that there were few to no plastic beverage bottles on the shelves in grocery stores, and certainly [almost] no one drank water that was purchased. Glass beverage containers with a "return" value were standard. Since my childhood there has been an explosion of applications in the use of plastic, not only in the food and beverage industries, but everything from wide use in auto production to computers.

Within the last few years great strides have been made in the reduction of the amount of plastic used to produce food and beverage containers. An empty half-liter water bottle has been reduced from 22 grams to 8.5 grams. (Soda bottles, in order to withstand the pressure of carbonation, however, have not.) Similar reductions in the weight of aluminum cans has occurred. Still plastic and aluminum waste is enormous. Each year (based on a 2008 report), as already stated 29 billion plastic water bottles are produced for use in the United States alone, according to the Earth Policy Institute, an environmental organization in Washington, D.C.[16] Manufacturing these bottles requires the equivalent of 17 million barrels of crude oil, yet even the escalating price of oil and therefore the cost of plastic has done little to curb the demand for plastic containers and other uses. Clearly we must ask ourselves: "If aluminum and glass are more environmentally friendly alternatives, why are we using so much plastic?"

The answer appears to be in the economies and the markets within which these products and the containers they require. All three beverage container manufacturers of glass, aluminum, and plastic tout their beneficial recycling rates. Plastic manufactures claim the advantage of cost, weight and condensed packaging. Glass and aluminum can manufacturers claim endless recyclability. Aluminum probably has the environmental-friendly advantage in that making a can from an old can instead of the raw material uses 5 percent of the energy and generates five percent of the emissions; a claim that cannot be made by the others. Also, cans keep out light that can degrade a product and lead to waste. Within the perspective of aluminum cans, reuse and recycle become basically one concept. Further, unlike glass and plastic, the value of aluminum makes it far less likely to be discarded since it is far more likely to be picked up by someone for its value when turned in.

Reducing Consumption

It is the natural progression of human behavior to make our lives easier. Naturally we want to shelter ourselves from the extremes of nature in temperature and weather. There is a reason the cave man lived in caves. He wanted to make his existence more comfortable. For most of us who have experienced electrical power failures in our homes, it is uncomfortable. But we consume electricity and energy voraciously. Can we not in the interest of conservation get by on less? Unless the government steps in to regulate our consumption, the only other regulation beyond personal conviction is the cost of this energy.

Isn't energy use, after all, a personal choice? Who is to say what level of energy consumption anyone should live within? To some, major consumption reduction might be low energy lighting, unplugging unused cell phone chargers, or less aggressive thermostat settings. To others it may be fewer trips in the family jet. Consumption is and always will be primarily related to and regulated by levels of affluence and prosperity. Pick your pleasure, but generally that pleasure will always come under the criticism of excessive energy consumption.

Conclusion

The cap and trade legislation failed to reach President Obama's desk when it died in Congress mostly because the public gave their representatives strong feedback that cap and trade would amount to a massive energy tax. President Obama then said "there is more than one way of skinning a cat." This other way was the Clean Power Plan. This plan which was finalized in 2015 would have established the nation's first CO_2 emissions standards. It drew criticism because unelected officials were now poised to do what Congress would not do: impose higher energy costs on Americans for what they believed to be questionable methods to reduce perceived global warming. In the end, it was not implemented having been stayed by the U.S. Supreme court, in February 2016. President Trump revised much of the plan after being elected, calling it the Affordable Clean Energy (ACE) plan.[17] This plan is much more coal-friendly and places more authority with state regulators and promotes clean-burning coal plants. Its goal is to decrease emissions by far less than the Clean Power Plan, but overall would reduce electricity prices. At the time of this writing, the ACE plan has not yet been implemented, still being debated in Congress.

Recycling benefits the environment when energy use is reduced below the amount needed to manufacture those products. Reusing, for the most part, is far more energy efficient and environmentally friendly. Reducing the use of plastic containers is more environmentally friendly and should be an immediate priority in cleaning up our environment and reducing energy consumption.

It is easy to proclaim the stresses on our environment would be reduced if we all consumed less. Our consumption, however, is based on economics. The more affluent will consume more, emit more, and waste more. If we really want to cut consumption, there is much we can do in so many areas to be better stewards of our planet. It all begins with becoming active and proactive in our communities and in our own lives:

- Worldwide environmental stewardship needs to be part of the basic education of our children.
- We need to wean ourselves off of using plastic containers.
- Use reusable containers for retail shopping. When the cashier asks: "Paper or plastic?" Our response should be: "I have my own, thank you."
- Don't buy beverages in plastic containers
- Recycle all aluminum
- Reuse first, then recycle, then landfill.
- Though cap and trade sounds like a possible way to control carbon (or any other) GHG emissions, energy consumption remains a personal choice and its regulation would require a large government presence in our lives to implement. It is closely linked with the development of the renewable energies discussed previously and again later in chapter 11. We will be better served as caretakers of our environment at this time by concentrating on "clean up" rather than GHG emissions until science can determine the quantifiable affect anthropogenic GHG emissions have on climate change and ultimately global warming.

CHAPTER EIGHT

SUSTAINABILITY

"Sustainable development meets the needs of the present without compromising the ability of future generations to meet their own needs."
—Brundtland Commission of the United Nations

Introduction

Sustainability is at the heart of the environmental movement. The concern is we humans are consuming natural resources at rates that are not sustainable into the future while polluting the planet and its atmosphere at destructive rates with the byproducts of that consumption. From this concept come the terms, "carbon footprint," "carbon tax," "cap and trade," and "Green Energy" to name just a few. The concept of sustainability, as it has been used within the environmental community, has its origins in the 1960s[1] as a product of the social revolution that began during that turbulent decade. Concerned with rapidly increasing human population, Stanford University Professor Paul Ehrlich wrote a book *The Population Bomb* which expressed alarm about the planet's ability to sustain human habitation due to this growth and increased consumption habits of humans (Ehrlich 1968). His warning became the focus of both concerned environmental and social activists. At its core is the belief that we humans are consumers on a scale that threatens the planet's ability to "replenish" or "sustain" itself in the midst of these increasing consumption demands.

"Sustainability" has become the new catch word for concerned environmentalists and for the green movement with multiple meanings impacting human consumption activity. Sustainability means: not depleting natural resources in order to maintain long-term ecological balance, development that meets the needs of the present without compromising the ability of future generations to meet their own needs, and doing no harm to the environment. The concern is we humans are consuming natural resources at rates that are not sustainable into the future while polluting the planet and its atmosphere at destructive rates with the byproducts of that consumption. This pollution further reduces our planet's ability to maintain its ecological balance. Those concerned with sustainability make two assumptions: 1) that most, if not all, Earthly resources are finite, and 2) everything that's no longer needed should eventually be processed into something else with neutral to little energy added to subsequent production. Central to their belief in creating a sustainable world is that our energy needs would be powered by renewable energy and waste would be very close to zero creating a system that would be a lot more sustainable than the polluting, wasteful, and throwaway system of production and consumption we now have.

Discussion

To understand this concept we can use the example of nomadic tribes throughout history that sustained themselves in one area while consuming all the resources of food, fuel, and shelter needs that were available and then had to move on in search of more of those resources. Unrestrained use and consumption of those resources led directly for their need to be nomadic as each new habitat could no longer sustain their existence. Eventually, as populations of nomads increased, their habitats began to overlap and they had to compete for scarce resources often time resulting in conflict and cultural changes.

With the transition from nomadic, to permanent farming settlements, to urban centers, sustainability took on a more collective mission. Consumers and polluters have become the main bad guys in this concept of sustainability. They are the CO_2 and sulfide emitters, the mercury dumpers, the strip miners, and the growers using fertilizers and mechanization, the manufacturers with their packaging and associated waste are using up resources and polluting at unsustainable rates. The automobile is singled out by those who steadfastly believe humans are the cause of global warming for its contribution to the increasing amounts of CO_2 being emitted into the atmosphere. They believe the automobile not only adds to the sustainability issues of the atmosphere, but also the immense land use required for their roads and interchanges which pressure the ability of the Earth as well, to sustain itself.

I would add the following example of the movement toward sustainability as an indication of how this concept would translate to a global effort. My alma mater, Macalester College in St. Paul, Minnesota, has stepped up with a pledge of sustainability. Their Sustainability Office is fast becoming one of the most well-staffed and well-funded offices on campus. With a Zero Waste Committee in addition to regular student employees the office has started on its campaign to make the school zero waste by 2020 and carbon neutral by 2025.[2] Such a policy comes with a cost. After a self-imposed printer awareness week, a strange initiative surfaced; students were informed that they would be required to give their user credentials (Macalester login) in order to perform any print project in the library. A maximum of three copies could be made at any one time. This policy change was made with no input from the students. As a result the Sustainability Office's policy becomes imposing and accountability becomes an issue.

The concept of sustainability is, of course, a noble one, espousing the "waste not, want not" philosophy. The hidden agenda within the concept is in its implementation. Since environmental issues are global, so must be the sustainability effort to be effective. Since the inhabitants of this planet all have different needs, goals, and different attitudes based on predominately culture, sustainability goals and requirements will vary across the planet. Ultimately those who so strongly advocate this concept are reluctant to admit or be upfront about the understanding that their concept of sustainability would require a world government to implement and control.

How much energy should be used by each individual to live their lives? How warm/cool should homes be maintained using which energy source? How many watts should light bulbs require for "adequate" lighting? How many miles per gallon should be required in vehicles, and what if some require trucking, flying, or shipping on the open seas to maintain their individual pursuits and livelihoods? How much should one be expected to "walk" to get to any destination? How should one vacation and what leisure pursuits would be considered "environmentally sustainable"? And what should we do about those pesky recreational vehicles? Should one buy liquids in plastic bottles or reusable glass containers, or recyclable aluminum containers? At the grocery store should one use "paper or plastic" or reusable cloth bags? Should produce and meats be wrapped or placed in sealed containers to cut down on waste? Should flush toilets only require three liters/flush or five, or should adjustable flush rates be used? I could continue in volumes to offer examples of this. The issues of sustainability dominate our entire lives and how we each choose to sustain our lifestyle and who should ultimately be accountable for maintaining a sustainable environment become highly personal and individualistic. Solving the issues of promoting sustainability are daunting, yet crucial to practicing good stewardship of the environment of our planet. Our efforts must start with awareness and then to education. Complicating the problem on a global level is that poverty is not conducive to promoting awareness and required education to implement good policy.

Conclusion

We humans have now inhabited nearly every bit of land mass in varying degrees on this planet and through technological advances in both transportation and communication we have been able to amazingly connect ourselves to the rest of the inhabitants across the vastness of the Earth. Along with this has come increased stress both on our environment and the human ability to sustain life as we know it on this "shrinking" planet and it available resources.

The discussion of sustainability is really a discussion of economics. We all deal with scarce resources every day of our lives. We manage them to sustain them, and with that management we prioritize those resources based on our individual needs and values. Technology also becomes a major influence in sustainability. New sources of energy and more efficient means of growing crops and managing the environment, for example, are always being developed. I remember the days of the Mid East Oil Embargo when our concerns were "we're going to run out of oil." Due to technologies in the oil and natural gas industry, some old and many more emerging, we now have more oil and gas then we ever thought was available to us. Nuclear "waste" still contains 98 percent[3] of the energy it originally had and is just waiting for the technology that certainly will be developed to capture it and help sustain our energy needs further into the future.

For a society to live and practice a sustainable lifestyle means "moving" toward a greener existence with humans emitting fewer polluting gases, growing their own food on smaller plots of land, creating less waste, and leaving less of a trace of humankind on our planet. To appeal to the masses and to be able to sell sustainability to the masses, greener should be better: cheaper to own, cheaper to use, healthier, higher performance, more durable, more stylish, more repairable and reusable, etc. I would add that greener shouldn't embrace "poorer in quality" in any large way.

The management of those resources globally will require some sort of global government. That political event is for another discussion.

CHAPTER NINE

GEOENGINEERING

"Geoengineering - the deliberate, large-scale manipulation of the Earth's climate to offset global warming - is a nightmare fix for climate change."
—Jeff Goodell, Writer, and Fellow at The New America Foundation

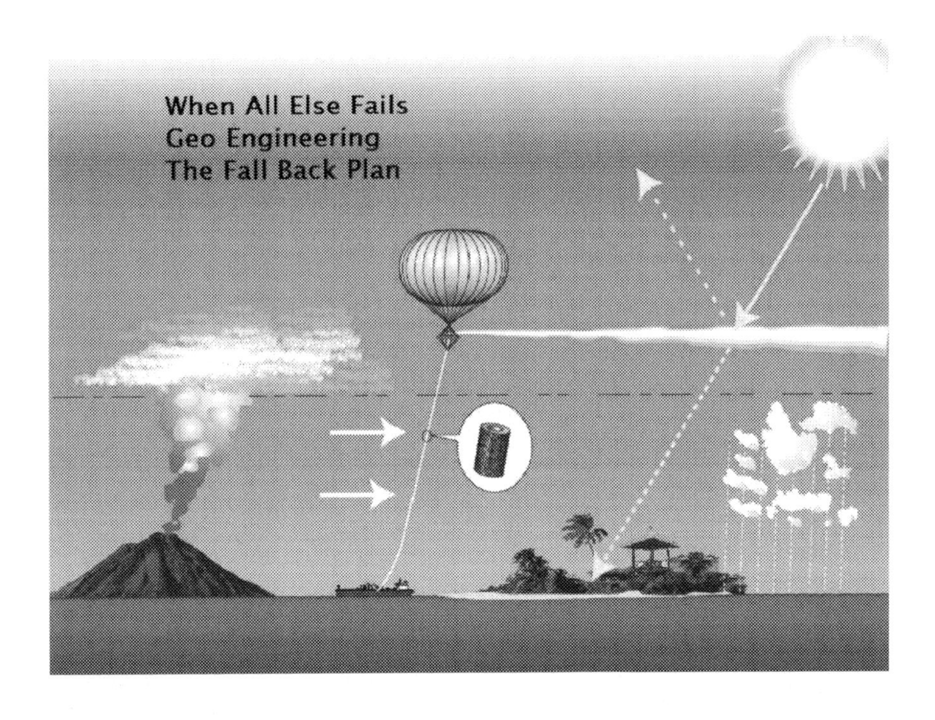

Introduction

As concerns grow that the planet is heating up and the primary culprit, CO_2, is blamed, there are emerging technologies being promoted by a group of scientists to mitigate the increasing concentration of this gas in our atmosphere. Some of their ideas at first appear to be quite "out there" and in the realm of very narrow and seemingly impossible realities for us common residents living on this vast orb spinning in space to comprehend.

Most have probably not heard of these technologies. Have you heard about putting mirrors in space to reflect the sun's energy back into space therefore cooling the planet, or technologies that that might be able to remove CO_2 and other greenhouse gases from the atmosphere? Have you heard about employing technologies that would mimic the cooling effect that large volcanic eruptions have had on the Earth's temperature and its climate that we have witnessed with past eruptions such as those at Mt. Pinatubo and Mt. Krakatau in the South Pacific? All these ideas and more belong to a diverse group of "**geoengineering**" technologies that have recently increased interest both in the scientific community and as well as among more than a few governments. I include this topic as it is being promoted by many scientists as a last alternative effort to avoid the tipping point in global warming.

Discussion

Geoengineering is the science of human-induced climate change and its objective is to cool the planet through deliberate large-scale intervention in the Earth's climate system. By slowing down the perceived recent rapid increases in global warming and to prevent a predicted "tipping point" beyond which temperatures will continue unabated to rise. The term 'geoengineering' relates to the various strategies and techniques aimed at containing and, in some cases, reversing the effects of both anthropogenic and known natural forms of observed global warming and environmental degradation. These strategies and techniques range from the fairly safe to the highly scientific and politically controversial.

Does this sound like a daunting undertaking? Yes, and as we shall see, it could create as many problems as it intends to resolve; if it proves to be possible. From a standpoint of environmental concern, geoengineering treats the symptom, it does not cure the apparent problem (increasing CO_2 emissions) and is being pursued merely as the fall back method for intervening in a potential planetary emergency.

There are two primary efforts in geoengineering. The first, **Carbon Dioxide Removal** (CDR) entails natural and engineered mechanical means of **sequestration** to reduce concentrations of CO_2 (and to a lesser degree, other greenhouse gases) through a process called "carbon capture." The second effort is called **solar radiation management** (SRM) the aim of which is to prevent solar energy from reaching the planet. This involves, for example, putting sulfur aerosols (which mimic volcanic eruptions and results in engineered "cooling") into the high reaches of the atmosphere using increased albedo technologies to reflect solar energy back into space. Geoengineers are working primarily on these two approaches to offer doable means to "engineer" our way out of human-induced global warming and any subsequent climate change results.

Regarding the first issue, high concentrations of greenhouse gases, the obvious and most certain way to reduce these is to reduce sources of man-made (anthropogenic) emissions. Such a decrease should be in the realm of possibility given that anthropogenic activity constitutes a small percentage of overall emissions, right? The United States has succeeded in reducing CO_2 emissions over the past few decades through the following means: increasing the use of natural gas; cleaning up much of its coal-fired energy generation facilities; increasing internal combustion engine efficiency; and improving others as well. In spite of U.S. efforts to curtail the anthropogenic generation of greenhouse gases, particularly in the last fifty to sixty years, Third World emissions have continued to rise worldwide. In this effort to reduce CO_2 emissions, it is much easier for economically advanced countries to "**green-up**" their infrastructure. Developing countries that lack necessary resources and technologies would have to achieve formidable shifts in economic functioning to realize the same results.

When I first started researching these ostensibly impractical means for reducing emissions, I was beyond skeptical. However, after following some of the leading scientists in this field (primarily those included in the geoengineering group of Ken Caldeira and Mike MacCracken), some of their reasoning began making sense. Geoengineering efforts are meant to be a short term fix or interim fix until anthropogenic green house gas production can be reduced back to what many believe to be more sustainable levels. My skepticism remains but is now centered on the magnitude of the issue, rather than the technical aspects. Our atmosphere is so immense that to have an impact, at least in the short term, seems infinitely impossible. There have been no large-scale geoengineering projects undertaken at the time of this writing, and its applicability has been through modeling, a method which at its best can only be considered incomplete.

In the case of CO_2 reduction, concentrations have increased from 250ppm a hundred years ago to 400ppm today[1]. Many scientists have claimed in the past that 400ppm could be the tipping point beyond which global warming trends could not "recover[2]." If it can be shown that CO_2 concentration levels could reach such a definable tipping point that would lead to an irreversible warming trend, what better way to deal with those levels than to "engineer" them away from such hazardous concentration levels?

Let's first look at some of the strategies to remove CO_2 and carbon from the atmosphere called carbon sequestration or capture.

Carbon Dioxide Removal

Biochar is a plant byproduct similar to charcoal that is made from lumber waste, dried corn stalks, and other plant residues. A process called "**pyrolysis**" heats the vegetation slowly without oxygen and produces carbon rich chunks of biochar that can be placed in the soil as a fertilizer, which locks the CO_2 underground instead of letting the CO_2 re-enter the atmosphere. Estimates of biochar half-life vary greatly from 10 years to more than 100 years. The effort here is to "pyrolysis" more plants which would lock away the carbon they absorb and then turn them into biochar, or charcoal, that can be mixed into soil. This plan is safe, and will slow CO_2 emissions and its resulting climate-changing effects. However, the overall effectiveness of this method, which would compete with food and biofuel production (as this resource would be in effect buried rather than consumed as food or fuel) is questionable as to the extent it could actually offset CO_2-induced warming.

Reforestation: As we saw in chapter 3, plants already convert atmospheric CO_2 into solid materials. Reducing the continued destruction of forests and encouraging the growth of new forests could tie up a lot of carbon in plant material.

Improved **soil management**: This is also a promising technology because it prevents the loss through tilling of carbon already sequestered in the soil. Leaving slash (plant waste left over after crop production) on fields to be incorporated into the soil reduces the carbon that would otherwise be emitted to the atmosphere.

Ocean fertilization: The more CO_2 the oceans absorb, the more acidic (actually "less alkaline" since ocean pH is 8 plus) they become. A geoengineering concept to absorb some of this increased CO_2 is called ocean fertilization. The plan is to fertilize the ocean with iron which would boost the growth of plankton. As the plankton grow and die they would take with them the carbon they have absorbed to the bottom of the ocean and be buried, safely locking it away. This long term approach plan is less intrusive to the atmosphere in reducing concentrations of CO_2. This will also slow the rate of ocean acidification as CO_2 is photosynthesized by the plankton and then the associated carbon is carried to the ocean floor where it is buried. This concept has already been tested unilaterally (unauthorized by any governing agency) by German researchers in 2009 during a five week period of fertilizing a thirty-seven mile-wide eddy they found in the Antarctic Ocean.[3] They scattered seven tons of commercial iron sulfate particles, which developed into a giant bloom of diatom plankton within four weeks. The diatoms then died, sinking in clumps of entangled cells, to depths exceeding 3,250 feet. This test came under much criticism by others in the scientific and environmental communities as it was conducted without accounting for the risks it may have posed to the environment of the ocean.[4]

Sodium Hydroxide: CDR chemical engineers have known for decades how to remove CO_2 from the atmosphere using sodium hydroxide, a caustic base commonly known to us as lye. It will bind with CO_2, to make carbonates, thus eliminating the gas. That's how CO_2 is removed from the air in submarines and spaceships. The "mechanical" method based on this procedure involves capturing carbon dioxide from ambient air, compressing it and storing it in geologic reservoirs (usually porous rock or depleted oil wells). Though not called "fracking," this injection of CO_2 into oil fields is a proven method to force out extra oil. Carbon Engineering, for example, is a Canadian Company that is attempting (with funding from Bill Gates) to commercialize this technology to directly capture CO_2 from the atmosphere.

If this sodium hydroxide technology is successfully developed and its costs fall low enough, the gas would have many customers, including the oil industry itself which would buy the gas to inject into oil fields to force out any extra oil that cannot be pumped through normal means. As Howard J. Herzog, a senior research engineer at the Massachusetts Institute of Technology explains, "The injection has minimal risk… The enhanced oil recovery industry has put tens of millions of tons of carbon dioxide into the ground every year for decades with no problems."[5] Currently much of the carbon dioxide now used for this type of recovery comes from naturally occurring underground reserves that are piped to the oil fields.

More potential uses for captured CO_2: Another potential use for captured CO_2 is that it can be combined with hydrogen to make gasoline or diesel fuels. Though this technology sounds "out there," its successful implementation could close the carbon cycle by allowing us to use hydrocarbons indefinitely and thus eventually replace oil. While this strategy is feasible, it is not currently considered viable because it entails prohibitively expensive long-term, large-scale operations. A pilot plant has yet to be built, so costs of this technology remain unknown.

Coal-fired energy: At the time of this writing, a first-of-its-kind coal-fired power plant retrofitted with technology that would capture and store much of the carbon dioxide has begun operations in Saskatchewan, Canada. A similar U.S. government-backed project in Meredosia, Illinois that tests the feasibility of capturing and storing carbon from a coal burning plant and demonstrates a cleaner way to burn the world's most abundant fossil fuel is also underway. However, its developers face many legal and financial constraints and it remains to be shown if this is a viable process.

In summary, these proposed CDR techniques to remove CO_2 from the atmosphere show promise, but require long term implementation, generally decades, before a significant impact on the global atmospheric CO_2 concentrations could possibly be achieved.

Solar Radiation Management: Promising, Wacky, and Entirely Untested

The advantage of this area of geoengineering is that compared with long-term carbon capture strategies it will more quickly affect cooling technologies. In the following partagraphs, I will briefly review geoengineering concepts by drawing on the summary of Michael Le Page presented in the September 20, 2012 *New Scientist*.[6]

We know that volcanic eruptions pump huge amounts of sulfur dioxide (SO_2) into the stratosphere and that the aerosol effect has a demonstrated cooling effect on the Earth's surface. Expanding on this insight, geo engineers, in one plan, would use a fleet of airplanes to pump sulfur dioxide into the stratosphere, where it would form aerosols that would reflect sunlight straight back into space thus cooling the planet. The up-side of this effort is the potential, to offset potential warming from the increases in anthropogenic CO_2 emissions by reducing the amount of solar radiation penetrating our atmosphere. This conceptual model has not yet been tested on a global scale, so its effectiveness, feasibility, and affordability are yet to be determined.

Another plan, called **marine cloud brightening** involves the use of a fleet of ships to spray sea water into the air. The salt particles should increase the concentration of droplets in clouds, making them whiter and more reflective, thus cooling the planet. This plan has all the benefits of SO_2 injection but wouldn't cause chemicals to be injected into the atmosphere. As with SO_2 injections, real world tests of the effectiveness, feasibility, and affordability of marine cloud brightening have never been conducted.

Dispersing cirrus is a plan which includes spraying bismuth tri-iodide high in the atmosphere to seed formation of large ice crystals that rapidly fall to the ground. While this approach should reduce the amount of heat-trapping cirrus clouds, it also remains untested.

The use of **space parasols** (I admit this one is "out there," and mention it only to show where geoengineering concepts could take us) is a plan to deploy millions of sunshades into space to shade the Earth from the sun. Again, this is only a conceptual idea being considered by the geo engineers and is certainly not achievable with existing technology.

Other down sides of the SRM strategy are significant. Rainfall would be reduced because of the increased sunshade, and these altered regional climates could produce unwanted effects such as famines. The poles, both Arctic and Antarctic, would not cool as much as the tropics because of sun angles therefore doing little to reduce any ice melt occurring in the polar ice sheets. SRM won't stop ocean acidification, and if CO_2 levels continued to rise (assuming CO_2 is the warming culprit), the planet would probably warm more rapidly if these SRM measures were for some reason stopped.

Recap/Analysis

At best, at this time, scientists and policy makers consider these two focus areas (**SRM** and **CDR**) of geoengineering as a "band-aid" fix to preventing a perceived catastrophic tipping point that would cause the temperature of our atmosphere to soar. The merit of these plans is that they represent a potential last-resort to cool the planet if, in fact, warming continues unabated. Naturally, extensive research is required to demonstrate this as a viable, and safe solution in the short term.

One advantage in pursuing this approach is its low cost in comparison with the costs of reducing emissions both in dollars and in quality of life (a matter I will address in chapter 10). Geoengineering at least holds the promise of addressing global warming concerns more economically, and would offer some "emergency back-up options" to address global warming by rewarding scientific innovation. Though this effort treats the "disease" rather than the root cause, from most perspectives, its potential benefits warrant further study and research.

There are also moral and social issues that must be addressed. In many regions of the Earth, global warming is a desired change (for example longer growing seasons, and lower heating costs). Who gets to set the thermostat? As with sustainability, this plan would require a world-governing body, and how could such commitments be made to such a governing body within the framework of any given government, especially any democratic government? Any coalitions that commit to geoengineering efforts or to reduce emissions would be incentivized to include more countries due to the economic principle of "economies of scale." However, countries would also have incentives not to participate, in order to avoid costs associated with emission reduction while benefiting from reductions made elsewhere. Such is always the dilemma of group and bureaucratic efforts.

Conclusion: The Possibilities and Pitfalls of CDR and SRM

Many scientists believe that the industrial revolution marked the beginning of the "anthropocene," the age defined by the beginning of human influence on the Earth's ecosystems. The ability for humankind to now engineer the climate seems at least possible in theory, and this ability inspires the debate on whether man with this possibility of being a "geological agent," can influence the sustainability of the planet for human habitation.

There is insufficient knowledge to quantify the extent to which CDR or SRM could offset CO_2 emissions on a century time scale, and efforts to mitigate global warming and climate change using these geoengineering concepts face numerous hurdles. Among them are technical feasibility, cost, ecological risk, public opinion, capacity to regulate, ethical concerns, and ultimately—who gets to control the thermostat. Modeling indicates that SRM methods, if practicable, have the potential to substantially offset a global temperature rise, but they would also modify the global water cycle. If SRM were terminated for any reason, there is high confidence among geoengineers that global surface temperatures would rise and possibly quite rapidly. CDR and SRM methods carry side effects and long-term consequences on a global scale.[7] Scientists acknowledge these technologies are in the very early stages of conceptualization, as well as development, and remain untested on a global scale.

For the purpose of making a point, let's assume these geoengineering concepts prove doable. What if I, as the ruler of a nation, began to feel the adverse effects of climate change, and unilaterally decide to start reflecting sunlight back into space? What if this has the effect of altering the rainfall, for example, in another nation as well? It is not difficult to see how quickly the "Cold War" logic of expected threats and counter-threats would creep into international politics and thus present a major obstacle to global geoengineering efforts.

The concept of geoengineering does at least offer a less intrusive and lower-risk strategy to complement emissions reductions potentially "winning the hearts and minds" of those asked to make changes in their "carbon footprint." Such is the "intervening" intent of geoengineering while the relatively slow anthropogenic changes are being made. Though the vastness of our planet and atmosphere make it highly difficult to conduct much needed scientific research, but it would still be wise to explore the potential of emerging technologies. Prudence dictates this technology be "kept alive" with continued research.

"What could possibly go wrong?" is the question that science and scientists must be able to answer in order to proceed with this approach to climate control. As Steve Rayner of the Oxford Geoengineering Programme astutely observes: "Throughout human history the technologies of one generation created problems for the next. We have to find some way to deal with that; it's part of the evolution of human society."[8] Taking the planet and its residents into the unknown with scientists and governments playing "God" is an enormous but perhaps unavoidable next step to be endeavored as judicially possible.

CHAPTER TEN

EVERYTHING HAS A PRICE

THE COSTS OF GOING GREEN AND THE ENVIRONMENTAL IMPACT

"Global Warming is a real phenomenon, it is mostly man-made, and it will have a long run overall negative impact. This says nothing, however, of the efficacy of government attempts to reduce temperature through international agreements, nor of the economic costs involved in implementing such promises. Yes, climate change is a problem, but it is not emphatically the end of the world"

—Bjorn Lomborg

Introduction

In Econ 101, we learned that "everything has a cost." Everything has a value based on scarcity of resources, yours and the Earth's. In our private lives we certainly know this as we pay for the things our lifestyle needs and enjoys, and we can afford. The cost can also be in doing without or sacrificing wants and desires as we adjust our financial budgets for more valued needs. I have mentioned the words of my father before: "If you want to understand the impact of an issue take it to the ridiculous." If we are concerned about CO_2 increasing concentrations because of our human need to burn fossil fuels, why not just totally stop burning them. That will absolutely rid the planet of anthropogenic global warming from CO_2. A ban on fossil fuels would result in no more internal combustion engines in automobiles. Unless you are fortunate to live on the grid of a nuclear electrical generating plant, there would be no way to manufacture windmills, solar panels, paved roads, or high rise buildings. You will no longer have the means to cool your home; your bicycle will be your only alternative to walking. There will be no farm machinery to grow crops unless drawn by a horse. You will have no way to charge your Tesla, your cell phone or power your computer. There will be no more mass distribution systems and therefore no Walmart and no Amazon. It is safe to say most everything you own was once on a truck being transported somewhere. You get the picture. Until the energy technologies of solar, wind and even nuclear catch up, life as we know it would cease to be recognizable or tolerable for most.

On the other extreme, let's imagine a life with no limitations or regulations on fossil fuel energy use, with governments that do not regulate but rather promote and subsidize the use of fossil fuel energy. We would give up the great strides we have made to clear the smog that has engulfed our cities from the particulates automobile exhaust and power generating plants have put into the skies above our cities. The picture again becomes as harsh and intolerable.

As social inhabitants of this Earth it seems more sensible to ease our way out of using fossil fuels as more environmentally friendly energy means become available and affordable. Who then, would regulate and govern that transition for society to the "greener" energy use; the free market or local, national, and international governances? Again we can see the enormity and complexity of such a transition. Those of us whose environmental views are driven by the side of the "**alarmists**" would lean more toward a mandated transition by governments with greater dictatorial power. Those of us more on the side of the "**skeptical**" view would rather let democratic action and market forces direct any such transition. Both the views of the alarmist and the skeptic have costs, both financial and environmental; with both still not able to quantify the anthropogenic impact on the concerns of a warming planet.

Discussion

If the reduction of carbon emissions is our goal, we are told the most effective way is through fuel conservation, increased energy efficiency, switching to alternative low-carbon fuels and transitioning to the green energy sources such as wind and solar. The cost of this effort requires individual sacrifice and will certainly require a significant economic restructuring of governments around the world if the effort is global. This of course, will be virtually impossible and fraught with extremely complex and difficult political and economic hurdles to overcome. Will the wealthiest amongst us participate in the individual sacrifice equally with less affluent amongst us? As the wealthy countries and people are expected to bear more of the financial burden, will whatever governing authority ask them to maintain their carbon emissions at rates equal to the less wealthy nations and as well? Governments, not to mention individuals, naturally resist being "told" to change the status quo in order to slow down "their" portion of the greenhouse gas emissions. This will always be viewed as "unfair" when compared with a similar sacrifice not shared by others. Few among us want to be told what kind of car to drive, where to set our thermostats, what to eat, who gets to fly in private jets and posh tour buses, as well who gets to live in the big houses. Individual pursuits of happiness get in the way of such directives to be sure.

One of the leading researchers on the cost of "going green" is Bjørn Lomborg. He is a Danish economist, author, and visiting professor at the Copenhagen Business School as well as President of the Copenhagen Consensus Center. He is former director of the Danish government's Environmental Assessment Institute (EAI) in Copenhagen. As director of the Copenhagen Consensus Center, he brings together many of the world's top economists, including seven Nobel Laureates, to set priorities for the world. The Copenhagen Consensus Center is ranked by the University of Pennsylvania as one of the world's top 25 environmental think tanks.[1]

His research centers around dealing with the economics of environmental policy and those policy impacts on the citizens of this planet in fighting global warming, and climate change. Dr. Lomborg focuses on employing the smartest solutions first. His aim is to estimate both the temperature impact and the economic cost of those policy decisions primarily that arise out of the any environmental treaties, the latest being the 2015 Paris Climate Summit.

"When environmentalists around the world tell the story of global warming, they cast it as humanity's greatest challenge. They also promise that it is a challenge that they can meet at low cost, while improving the world in countless other ways. We now know that is nonsense. Climate change has been portrayed as a huge catastrophe costing as much as 20 percent of world gross domestic product (GDP) (if ignored), though brave politicians could counter it at a cost of just 1 percent of GDP. The reality is just the opposite: We now know that the damage cost will be perhaps 2 percent of world GDP, whereas climate policies can end up costing more than 11 percent of GDP." (Lomborg 2014)

At the time of this writing he had the only peer-reviewed estimate study of the Paris Climate Summit which showed that adopting all the promises from 2016-2030 would reduce the IPCC-estimated temperature increase by the year 2100 by 0.05 degrees C (0.09 degrees F). Even if all the countries continue their promised reduction until the year 2100, the temperature increase would only be reduced by 0.17 degrees C. (Lomborg 2017). Though the United States pulled out of the Paris agreement because favorable terms could not be negotiated, focusing specifically on the U.S.

policies demonstrates a negligible impact. Even if the U.S. further implements the entire set of the original U.S. climate promises, it will reduce global temperatures by the year 2100 by just 0.031 degrees C.[2]

By using the best climate-economic model ensembles available to him, Dr. Lomborg "determined the estimated cost of implementing the Paris promises at approximately $924 billion in GDP per year by 2030. And that is only if the policies are implemented efficiently and effectively. Peer reviewed literature suggests the economic costs will likely double to almost $2 trillion per year. The cost for the U.S. climate promises alone is likely to range from $154 billion to $172 billion every year in lost GDP by 2025, again doubling if not implemented efficiently." (Lomborg 2017)

To put these costs into a personal perspective the Paris promise for the U.S. was to cut emissions by 80 percent and Europe 80-95 percent by 2050,[3] a clearly daunting task. Without an electrical generating capacity from alternative "green" energy sources (which now provide at most 17 percent of world energy consumption) to make up for such cutbacks, this promise has no chance being kept. Imagine without green energy to take up the slack, one would have to cut their electrical consumption by 80 percent as well fossil fuel consumption by the same amount. Do we really believe the citizens of the Earth would make that sacrifice, all with the stated purpose of reducing warming by 2100 by less than 1 degree C?

An analysis of the Kyoto Protocol reveals that every industrialized nation promised in 1992 to reduce their emissions to 1990 levels by 2000 (United Nations 1992) and almost every single **Organization for Economic**

Co-operation and Development (OECD) nation missed that target; most claiming the goal was no longer possible because the economy grew more rapidly than expected (Joby & Baker 1997). Kyoto was abandoned by the U.S., Russia, Japan, and Canada.[4] It is therefore reasonable to assume future commitments will meet the same fate, and up to the time of this writing, they have.

Based on the U.S. Energy Information Administration (April 2016), the following two figures show both worldwide and U.S. energy consumption by source:

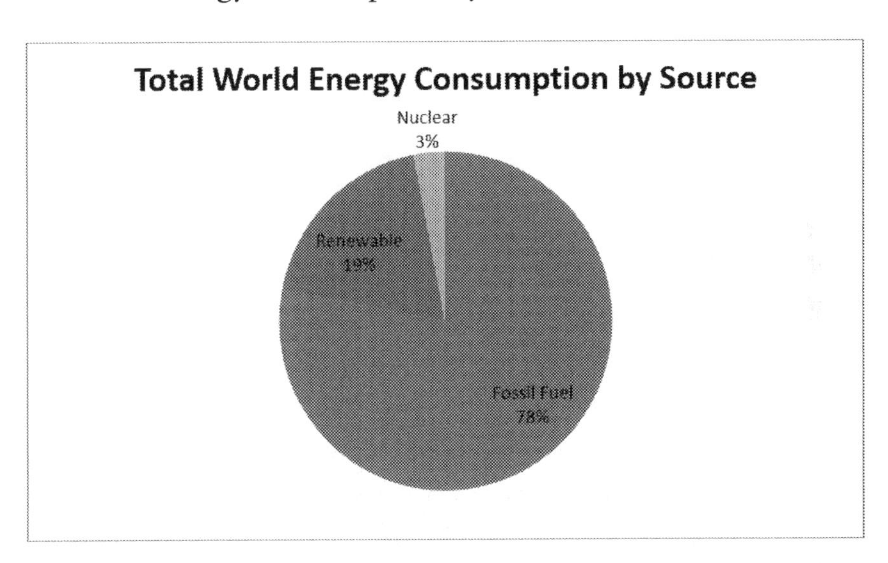

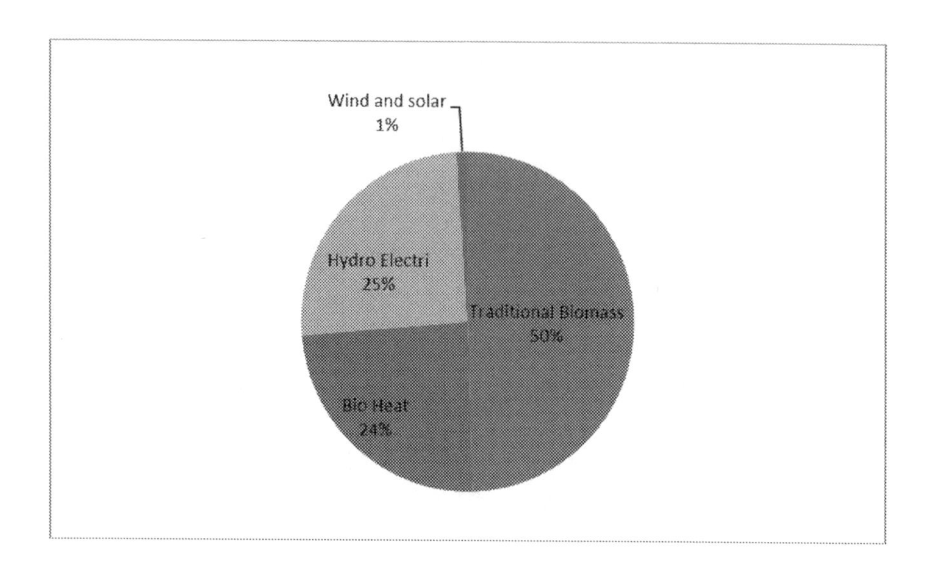

One more point is important to make which adds to the cost of emission reduction: the green energy industry is almost entirely reliant on government subsidy, and for the most part requires fossil fuel-burning infrastructure to produce and maintain it as a backup energy source.[5]

There are some sound practices of emission reduction that are low risk and relatively low cost that we can immediately employ that should be acceptable to most in the attempt to reduce CO_2 emissions. First, help nature sequester carbon through biological means. As we already know, plants convert atmospheric carbon into solid materials. Reducing the destruction of forests and promoting growth of new forests could tie up much carbon in plant material. Where deforestation is now economic opportunity for some third world countries, halting it would become a serious political issue for those economies while waiting for the development of alternative sources of affordable energy.

The cost of capturing the energy from wind is a good example of the cost of going green. In his book *Carbon Shift: How Peak Oil and Climate Crisis Will Change Canada (and Our Lives)*, Thomas Homer-Dixon, the Canadian political scientist, applies the concept of **net energy** to renewable sources of energy such as windmills and **photovoltaics** (solar). A two megawatt windmill (about average size for onshore windmills) "…contains 260 tons of steel requiring 170 tons of coking coal and 300 tons of iron ore all mined, transported, and produced by hydrocarbons. The question is: how long must a windmill generate energy before it creates more energy than it took to build it? At a good wind site, the energy payback day could be in three years or less; in a poor location, energy payback may be never and could spin until it falls apart and never generate as much energy as was invested in building it." (Homer-Dixon 2009)

Another example is the cost of solar power, an energy sector that has grown considerably in the last decade but heavily built by government subsidies paid for with tax dollars. Though the potential of this renewable energy source in cost savings is just being developed and its infrastructure costs falling, it is not a sustainable energy source without massive government subsidies. For example in terms of production, "subsidies for solar have received ten times the subsidies of all other forms of energy."[6] Solar energy especially when used commercially requires large amounts of land that come with a cost and unlike wind farms its does not generally share the land well with other uses. The impact of solar farms on the environment in land degradation, loss of habitat, interference with wildlife, wilderness and other protected areas will also have cost considerations.

Energy Comparison Costs

Germany can be used as an example of what the various types of electricity-producing energy sources cost, at least relative to each other. The prices of each of the technologies are fixed by law, and called a **feed in tariff.** According to Dr. Patrick Moore (one of the founders of Greenpeace) in a TEDxVancouver talk in 2009,[7] brown coal was 2.4 cents (€ per kilowatt hour), nuclear 2.5 cents, natural gas 3.0 cents, wind 10 cents, and solar 57 cents. Germany was, at that time, paying $9 billion dollars every year for solar energy and was getting less than 1 percent of their electricity from it.

Conclusion

Though some would have us believe that the inevitable transition to renewable energy sources such as hydro, solar, wind, and nuclear from the current GHG emitting energy sources will be affordable, for sure in the short term of the next 100 years, at least, it will not. The cost of going green will cause major changes in asset and resource allocation, and become a burden which will be borne with great sacrifice. What Dr. Lomborg has shown is that considering even the most optimistic reductions in GHG emissions, the temperature impact will do little to stabilize the Earth's climate. Since the result will only be to reduce temperature rise over the next hundred years by a fraction of a degree, suggests it would be a far wiser investment in the means to adapt to those small temperature increases. I will discuss the concept of adaptation in chapter 15.

CHAPTER ELEVEN

THE POLITICS OF GLOBAL WARMING AND CLIMATE CHANGE

"Climate change is the greatest challenge to our generation."
—John Kerry, U.S. Secretary of State

Introduction

In this chapter we will look at how the views of the warmist, skeptic, and denier may be formed and why. While each viewpoint claims it has the support of science, we can see (when not wearing those "blinders") from such diverse interpretations that the science of climate is anything but settled. It is relatively easy to agree on the facts, but few scientists agree on what those facts mean. I said I would keep politics out of this debate, however, in this chapter we'll look into how political viewpoints affect the environmental camp with which one identifies.

An important point to be made here is understanding how the statement is made. The assertion that global warming and climate change "are real" is fairly well accepted by all three camps. When this assertion is made it tends to draw all into agreement, however, all too often the asserter then assumes, without declaring specifically, that the warming and climate changes are caused by human activity. There is a big difference in the science between agreeing that global warming and climate change are occurring and then placing human activity to blame for those changes. The degree to which human activity contributes to global warming and climate change is what divides the warmist, the skeptic, and the denier.

I would suggest that political beliefs are a major factor on how one's opinion on global warming and climate change is based; specifically how human involvement affects the heating and cooling of our planet. A 2018 survey and report by Yale University uncovers some interesting beliefs and attitudes on global warming as it pertains to one's political viewpoint.[1] Drawing on a nationally representative survey of 1278 voting age adults, 1067 of which were registered voters; overall 59 percent of the respondents thought global warming was caused mostly by human activity. Of liberal Democrats, 84 percent believed the cause is human, but only 26 percent of conservative Republicans believed the cause is human. A 2016 Pew Research Center study indicated that the divide between Liberals and Conservatives on global warming and climate is due in part to trust issues.[2] Conservatives tend to mistrust the warmist scientists because they believe their research is more focused on promoting their careers rather than in a concern for public interest. They point to the scientists own political leanings and believe their careers would be in jeopardy if they go against their scientific community. Conservatives also fear that the warmists are using the environmental issue as a "Trojan Horse" to press their progressive/socialist agenda through the redistribution of wealth that comes with government control and leads to more regulation and taxation. Conservatives make the point there is plenty of grant money available from the government for the warmist's research and basically none for any research into the possible impact of natural variation, which they tend to support. Liberals tend not to trust the skeptical scientists because their research is based on a desire to help their industries which in turn drive social and economic inequity. The Liberal accuses the conservative of plugging their ears ignoring the available facts of what they believe to be settled science. Both sides have plenty of examples, which they will most often refer to as "facts" to support their claims.

Another reason for the divide politically is that Liberals believe government is required to "manage" the cure for most social and fiscal issues from poverty, social justice, to those of global warming and climate change. The conservative believes government involvement is generally costly, inefficient, and ineffective because bureaucracies would provide directives more supporting a political agenda rather than sound science. The conservative believes when a cure is needed, the free market forces will be much more efficient in both determining the cure and implementing it.

Another point of contention on the politics of perception and trust that is worthy of pointing out is the assertion that "there is a 97 percent consensus among scientists that global warming is human-caused." According to the National

Academy of Sciences, that consensus refers to: 1) the Earth is getting warmer, 2) the warming is mostly due to human activity, and 3) if GHG emissions continue, the warming will accelerate.

John Cook founder of the blog site *Skeptical Science* along with several other of his colleagues published a paper claiming they reviewed 12,000 "abstracts" of studies published in peer-reviewed climate literature.[3] Cook reported that he and his colleagues found that 97 percent of the papers that expressed a position on human-caused global warming "endorsed the consensus position that humans are causing global warming." The question Cook and his alarmist colleagues surveyed was simply whether humans have caused "some" global warming. The question is meaningless regarding the global warming debate because most skeptics as well as most alarmists believe humans have caused "some" global warming. The issue of contention dividing alarmists and skeptics is whether humans are causing global warming of such negative severity as to constitute a crisis demanding resolute action. Though global warming alarmists and many in the media have been reporting that the Cook study shows a 97 percent consensus that humans are causing a global warming crisis, clearly it was not the question surveyed. Once again there is a big difference in the way the assertion is made and consequently the way any conclusions should be drawn. An interesting point needs to be made here: investigative journalists found that Cook and his colleagues remarkably classified many papers by prominent and vigorous skeptics such as Willie Soon, Nir Shaviv, and others as supporting the 97 percent consensus,[4] but who have actually objected openly to the misrepresentation made of their work by John Cook.

The argument that many warmist scientists use to counter the position of those scientists who are skeptics or deniers is generally that they had significantly fewer publications suggesting they had done a lot less research.[5] Another claim is that the skeptical scientist is more likely a retired scientist making the inference that they are not as well-informed. It could also be said that a retired scientist is not so interested in his career and can therefore be more objective. And so the back and forth continues.

Despite the 97 percent consensus claim reported heavily in the media and elsewhere, a 2018 gallop poll found that 54 percent of Americans do not believe global warming will cause major problems in their lifetimes.[6] Polling throughout the world suggests about the same numbers, however, the trend of increasing concerns in the past of anthropogenic causes have begun to level off.

Another issue of trust which has done harm to climate science occurred in 2009 before the **Copenhagen Summit** when a server at the Climate Research Unit (CRU) at the University of East Anglia was hacked by an external attacker resulting leaked emails; an event which became known as "Climategate." What the leaked emails revealed was a group of the world's most influential climatologists arguing, brainstorming, and plotting together to enforce what amounts to a party line on climate change. Data that did not support their assumptions about global warming were "fudged." Peer-reviewed journals that dared to publish contrarian articles were threatened with boycotts. "When dissenting scientists sought freedom of information requests, the relevant emails were deleted and worse, original data was also likely deleted." (Lomborg 2010) The attempt of those emails to suppress critics suggested to skeptics that climate data and had been a long-occurring conspiracy to hide the decline in global temperatures observed in the first two decades of the twenty-first century. The CRU rejected the notion of conspiracy contending that the emails had been taken out of context and merely reflected an honest exchange of ideas.

Swedish climate scientist Lennart Bengtsson submitted a paper for peer review that suggested that climate is probably less sensitive to greenhouse emissions such as CO_2 than is acknowledged by the UNs IPCC, indicating along with his co-authors that more research is necessary to "reduce the underlying uncertainty."[7] He made this statement within the context that the IPCC has not exactly been a beacon of scientific certainty pointing to IPCC claims that islands would be disappearing under rising seas, (which they did not) and reports that the Himalayan glaciers are disappearing (which they are not). His paper was submitted but failed the peer-review process.

Dr. Bengtsson, who is a research fellow at the University of Reading and a member of the advisory council of the Global Warming Policy Foundation, was surprised by the rejection but also by the reaction of his peers that he would question existing climate change orthodoxy. He ultimately resigned from the foundation after being subjected to what he considered verbal abuse from a community he once respected. "I had not expected such an enormous worldwide pressure put at me from a community that I have been close to all my active life…it was 'utterly unacceptable' to advise against publishing a paper on political grounds"[8] he wrote in his resignation. He called it "an indication of how science is gradually being influenced by political views." One German physicist compared Bengtsson's joining the group to joining the Ku Klux Klan.[9] Such a claim by a scientist and criticisms from others can hardly be examples of science, but certainly exemplify the corruption of science by the presence of politics. I believe the warmist's lack of willingness to debate while obviously trying to silence opposing views in a scientific discussion, makes their position almost one of a religion and does not serve them well.

Another scientist, Professor Ivar Giaver, a U.S.-based Norwegian physicist, who is the chief technology officer at Applied Biophysics Inc. resigned from the American Physical Society (APS) which is the second largest organization of physicists. His peers there had elected him a fellow in honor of his work. This society which has over 48,000 members, adopted a policy statement which states: "The evidence is incontrovertible: global warming is occurring." In his resignation statement he told to the *Sunday Telegraph* he said "**Incontrovertible**" is not a scientific word. Nothing is incontrovertible in science" and in parting company with the society stated he could not live with its official statement on global warming.[10]

A leading supporter of then Senator Barrack Obama in his campaign for President, Professor Giaever has since criticized President Obama's position on global warming and was one of more than 100 scientists who wrote an open letter to him declaring: "We maintain that the case for alarm regarding climate change is grossly overstated." and questioned "whether the average temperature of the whole earth for a whole year can be accurately measured," and contended that "even if the results are accurate, they indicate the climate has actually been "amazingly stable" for 150 years."[11]

A discussion of the "involvement" of politics in the science of global warming and climate change would not be complete without mention of former Vice President, Al Gore. No single person has done more to bring awareness to the issues of global warming and climate change than he has. In his many books and an Oscar-winning award documentary on one of them (*An Inconvenient Truth)* he explains how humans have disrupted the climate of the Earth mainly through the human emissions of CO_2 from primarily the automobile and coal-fired power plants. He has declared the automobile and in particular the sport utility vehicle (SUV) the biggest contributor to the cause of global warming. (Gore 2006)

Al Gore bases his views on ice core tests over the last 650,000 years. He claims atmospheric CO_2 concentrations over the past 650,000 years as never exceeded 300ppm. As far as I can tell, he has not documented the source of that statement. From those same ice core samples, he says temperature can also be measured. While admitting that the relationship between CO_2 and temperature is complicated, his graph of temperature highs and lows over time matches up almost identically with the highs and lows of CO_2 concentrations. His point, "when there is more CO_2 in the atmosphere, the temperature gets warmer because it traps more heat inside [the atmosphere]." He points out that CO_2 is now at its highest ever at 350ppm (it is over 400ppm at the time of this writing). Nir

Shaviv an astrophysics, climate scientist, and physics professor at the Hebrew University of Jerusalem, responded to Al Gore's presentation in *An Inconvenient Truth*; "we have had 3 times and as much as 10 times as much CO_2 in the atmosphere as we have today, if CO_2 has a large effect on climate then we should [have] see it in the temperature reconstruction."[12]

Though the preponderance of scientists believe human activity contributes significantly to global warming, there are scores of other scientists, including Syun-ichi-Akasofu, Tim Ball, Judith Curry, Ivar Glaever, Richard Sanger, Frederick Singer, and John Christie to mention a few who believe the anthropogenic causes of global warming are unknown or are primarily caused by natural processes. Many other scientists believe there is no evidence CO_2 drives global temperature and therefore take the position that human causes of warming contribute only a small amount compared to that of natural variation. This contingent of scientists will be presented in a later chapter. Clearly the science of climate change and the causes of global warming are not a "settled" science.

Mr. Gore and other politicians have declared global warming a greater threat to our country than terrorism. The easy criticism of Al Gore's threat assessment, the spokesman for global warming and climate change, are his many predictions which have not come true. From his 2006 predictions of an "ice-free Arctic by 2013, 20 foot sea level rises in the near future, an increase in the number and severity of tornadoes and hurricanes," to his predicted "polar bear extinction," and that a planetary emergency will occur within a decade," (Gore 2006) which he made in his book *An Inconvenient Truth*, 12 years later none of them have come true. My issue with former Vice President Gore is that he "talks the talk" but does not "walk the walk." He leaves his speeches and conferences in SUVs to fly private jets on to his next environmental engagement while living in a mansion that in 2016 consumed over 230,000 kilowatt hours when the average family home consumes about 11,000 kilowatt hours. And while Mr. Gore has publicly said his home is completely powered by renewable energy, only about 3 percent of the electricity used in his Nashville residence comes from renewable energy sources.[13] He also does not tolerate nor debate those with opposing points of view. He was one of the first to declare the "science is settled, the debate is over."

Conclusion

It is extremely difficult and a robust task to separate politics from the environmental debate. One of the problems is that one side, that of the warmists, who believe human activity is the primary cause of climate change and global warming has shown an unwillingness to publicly debate, choosing it seems to hide behind the mantle of "the debate is over the science is settled."[14] I find it very difficult to believe a scientist would ever take this position; a politician, yes, but not a scientist.

To take some liberty with Thomas Wolfe's quote, "The essence of science [belief] is doubt, and the essence of reality is questioning." Why would a scientist or politician want to close down the debate in a seeming attempt to silence the opposition while claiming science is on their side? It appears to me that the skeptic is eager to debate and engage the opposing points of view. This should speak volumes as to which side of the issue is more interested in the science, does it not?

Until science can better quantify the human contribution to global warming and tell us the environmental path we should follow, shouldn't the politician stand back and not be so eager to enact public policy changes based on incomplete and possibly erroneous information? Science cannot do its job when debate is stifled.

CHAPTER TWELVE

OVERPOPULATION

"Overconsumption and overpopulation underlie every environmental problem we face today."
—Jacques-Ives Cousteau

Introduction

Is there a maximum number of human beings whose lives can be sustained by this planet? Certainly, there must be a finite number. The economics of this issue would suggest that the Earth's resources are finite, that we all live on a finite amount of terra firma, so it just seems there must be a sustainable limit to the human population of the Earth. On this global population density map[1] it can be seen where the human population footprint is concentrated and where it is not. Surprisingly, with a global population of seven billion people in 2015, the planet still remains "wide open" for habitation throughout much of its area, with the western hemisphere appearing to average fewer than ten people per square mile.

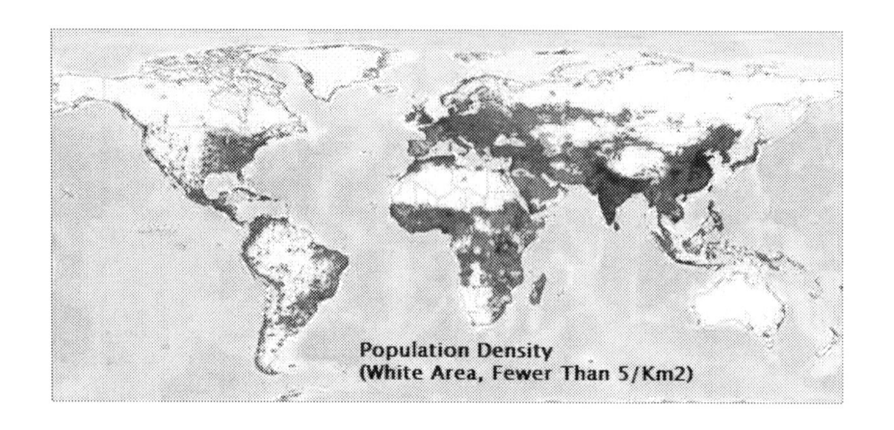

Discussion

When we talk about a globally sustainable population, what we are really talking about is "consumption" and its sustainable limits within scarce natural resources. In my lifetime I have seen the Earth's population grow from two billion souls to currently just over seven billion with a projected growth to over ten billion in 2050 according to the U.S. census bureau.

To some, this growth rate cannot be sustained into the future. To others this growth rate is "troublesome;" and still to others, of "little concern." Perspective is much of the issue. If you live in Tokyo, the most populated city at 33 million, New York City and San Paulo, with 18 million; or any of the remaining top 125 cities populated by over one million residents; you live in a crowd. If you live in Mubai, India, with 75,000 residents /square mile (29,000 residents /sqKm), or Kolkata, India, 62,000 residents / square mile (24,000 residents/sqKm); you live in a "densely packed" crowd.[2] If you live in Alice Springs, Australia, you live in an area with a population density of fewer than 0.1 people per sqKm. The perception of "overpopulation" and its impact on this planet and sustainability is vastly different depending on where one lives, as well as the life-style one chooses to live. So I would conclude that the concept of overpopulation has more to do with perspective than a number. The good news, however, the global population growth has slowed and according to some estimates the population of the Earth will stop growing within the lifespan of people alive today.[3]

As can be imagined the environmental concerns in those top 125 cities with more than a million in population are vastly different than much smaller cities, towns, and rural communities. It would be virtually impossible to get any kind of agreement or consensus among these culturally diverse residents of these population centers on environmental concerns in addressing any "overpopulation" issues. While the "leave no trace" environmental policy in sparsely populated communities is a viable concern and goal, it is not a viable or even doable policy in large cities and population centers. The "trace" in these areas is an undeniable scaring of the Earth with human habitat and the infrastructure required to sustain it.

To understand how these densely populated cities can sustain themselves requires an understanding of "Rural to Urban" **demographics**. This demographic has been a rapidly changing one since the beginning of the twentieth century with the global proportion of urban population rising from 13 percent in 1900, to 29 percent in 1950, to 49 percent in 2005.[4] According to the Global-Rural Mapping Project, the Earth's population passed this demographic milestone in May 2007 and became more urban than rural.[5] This is only a symbolic date, however, calculated from an estimation. By comparison, in the United States, the tipping point from a majority rural to a majority urban population came early in the second decade of the twentieth century. Today, 21 percent of the United States is rural (some states, however; Maine, Mississippi, Vermont, and West Virginia are still more rural than urban).[6] This graph shows the development of world demographics of urban to rural since 1950, projected to 2030.[7]

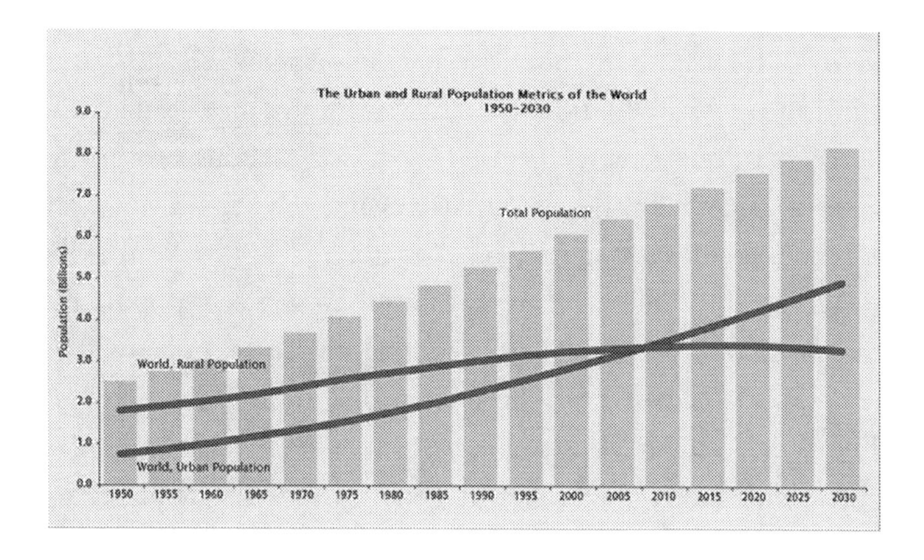

How does the population in urban areas like New York City and Tokyo, for example, feed itself? How does it sustain its supply of food? Despite what some may think, private gardens grown by those in the urban areas can do little to sustain the urban lifestyle choice, but can only augment the required food supply in a "feel good" way. So how is the vast requirement for food in urban "sprawls" sustained? Well, of course, it is produced in the vast rural areas of a very fertile planet.

The issue of sustainability goes well beyond the food supply, to include: clean air and water, waste treatment, transportation, recreation, etc., to mention just some of our concerns in densely populated areas. However, the critical issue that drives all the others is economic sustainability. While this urbanization demographic moved wealth from rural areas to urban centers, the urban areas still depend, for the most part, on the rural population to sustain its basic needs, particularly for food and clean water. Complicating this rural to urban demographic shift, is the environmental impact on rural areas as they also must deal with the urban waste products of polluted air, contaminated water, solid and hazardous wastes discharged and dispersed by those urban populations.

Further, global poverty remains concentrated more in the rural areas along with inferior educational opportunities. Because of this demographic shift, the political power has shifted from those who produce the food, and resources to those who consume them. The political consequences of this urban/rural demographic shift and these few noted examples are beyond the scope of this book but must be considered in any related environmental public policies and regulations imposed by governments on their citizens; locally as well as globally.

In the last 200 years the population explosion of 1.5 to almost 8 billion has been a quite sustainable growth, which many did not predict. In 1820 the vast majority of the global population lived in extreme poverty with only very few rising above that poverty standard of living. Economic growth has completely transformed our world as poverty and hunger have fallen continuously over those two hundred years. This is quite remarkable; actually, considering a population explosion of seven times occurred during those two hundred years. This more than sustainable population increase is due primarily to higher living standards and decreasing mortality rates. According to the researchers at "Our World Data" and specifically author Max Roser:

"In a world without economic growth, an increase in the population would result in less and less income for everyone, and a seven-fold increase would have surely resulted in a world in which everyone is extremely poor. Yet, the exact opposite happened. In a time of unprecedented population growth we managed to lift more and more people out of poverty! Even in 1981 more than 50 percent of the world population lived in absolute poverty this is now down to about 14 percent. This is still a large number of people, but the change is happening incredibly fast. For our present world, the data tells us that poverty is now falling more quickly than ever before in world history."[8]

Based on "Our World In Data,"[9] the percentage of the world living in poverty continues to fall in the twenty-first century.

Share of World Living in Poverty
(< $1.90 per day)

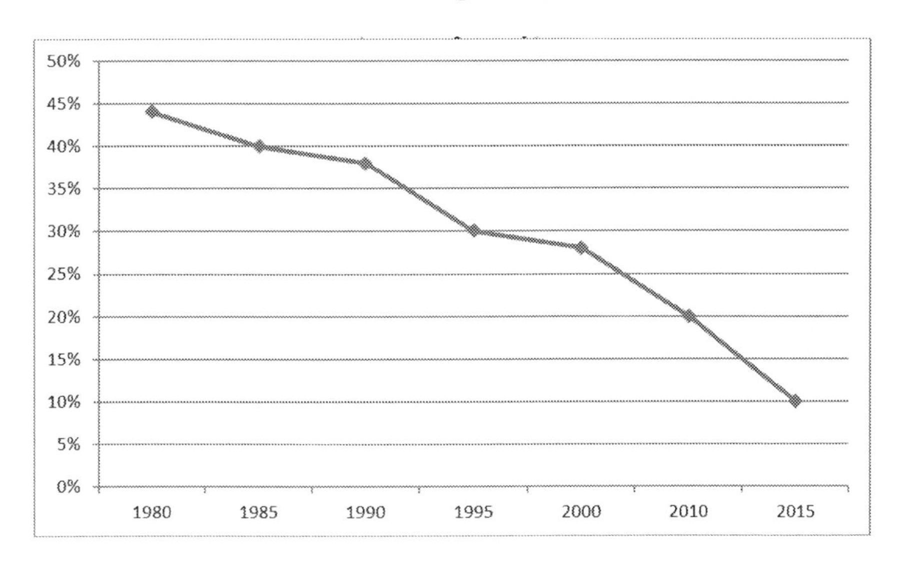

What is the reason for the reduction in global poverty with such population growth rates? In a word, it is "technology." Advances and improved production methods have resulted in huge increases in food and energy production, and innovation methods of transportation to bring needed resources to a growing population. Better construction efficiencies in housing, and medical advancements in curing, containing and identifying diseases has become even more possible with the population of the planet increasing by five times that of fifty years ago. So far the vast, sprawling, and ever-increasing urban centers can still get whatever resources needed and can be sustained by the increasingly more efficient production in the rural areas. Exceptions, however, abound. One only has to look at the country of India to see how rural areas that lack efficient transportation infrastructure cannot get much of their crops to market before they spoil. This dilemma is repeated in areas primarily lacking in education and the availability of cheap energy.

Cities must still refine and process rural goods for urban and rural consumers. But if either cities or rural areas had to sustain themselves without the other, few would bet on the cities. So both demographics remain dependent on each other. Dr. Ron Wimberley, Professor of Sociology at NC State advises "…given global rural poverty, the rural-urban question for the future is not just what rural people and places can do for the world's new urban majority; rather, what can the urban majority do for poor rural people and the resources upon which cities depend for existence? The sustainable future of the new urban world may well depend upon the answer."[10] However, this does not address the question of how many human beings can be sustained on the finite land mass of the planet. More humans, means more consumption and depletion of the "finite" resources of the planet. To understand this concept and the concerns of "overpopulation" a look back in history may help.

As was presented in chapter 8, in 1968, Stanford population biologist, Paul Ehrlich, published his gloom and doom book, *The Population Bomb.* In it he railed against consumption as a crime against humanity. He is well known for his 1968 predictions of global famine in the 1970s and 1980s where hundreds of millions of people, would die in great famines. Ultimately, of course, that did not occur; in fact quite the opposite occurred as the Earth's population has almost tripled since then. The Earth's population in 1968 was more than 3.5 billion already way over Ehrlich's goal of maintaining a "sustainable" global population around 1 billion, well past the point of even having a chance of sustaining itself according to Ehrlich. (Ehrlich 1968) According to the UN World Health Organization (WHO) statistics, the number of people in the developing world who were undernourished in 1968 was estimated to be more than 900 million. By 2018, this number had fallen to 821 million. [11] Though recent findings show world hunger may be increasing slightly, still the world's population has almost doubled, while global hunger has been reduced.

Further, as far back as the eighteenth century the English economist Thomas Malthus (1766-1834) hypothesized in his paper "An Essay on the Principle of Population," that while population would grow geometrically, food production would grow arithmetically leaving people with little to eat.[12] two hundred years later this catastrophe also has not occurred. Worldwide famine, though still a real problem in some areas, has in most areas been reduced. Those predictions failed to anticipate increased food production from the advancements of technology. We should anticipate the further development of technologies into the future to deal with population growth as well as new discoveries of resources and more efficient ways to use them. In many ways, we can safely say humans are just getting started adapting to population growth; and history suggests just that.

Another factor in population growth warrants discussion here. As the world's population prosperity and development proceeds, fertility rates will fall as has been shown in the developed world where live birth rates have fallen and slowed the rate of population growth in those developed countries. It is generally considered that a 2.1 live birth rate per woman is required to maintain population equilibrium. Much of the developed world has reached that birth rate.[13] Researchers at Austria's International Institute for Applied Systems Analysis foresee the global population maxing out at nine billion sometime around 2070.[14]

Conclusion

The maximum population the Earth can sustain cannot be known or determined. Advances in technology to produce more food, find more minerals, develop more resources, and to more efficiently use the finite land mass upon which we all must live; have more than kept up with population growth and the resource needs of an exponentially increasing population. Because poverty has decreased with this advance in technology, along with a decline in birth rates that go hand in hand with better technology, the concept of overpopulation is more rooted in perception than a real environmental concern for at least the foreseeable future. Further, we have not even begun to tap the potential of the oceans to support further population growth.

CHAPTER THIRTEEN

ADAPT/MITIGATE/ IGNORE?

"The measure of intelligence is the ability to change."
—Albert Einstein

"Those who cannot change their minds cannot change anything."
—George Bernard Shaw

Introduction

Humans may be considered to be the enemy of the health of the planet, but just as the destructive pine beetle is an integral part of the Earth's environment, so are we humans. Every living species on this planet, plant or animal, survives in varying degrees by transforming to the world around them. Due to our level of intelligence we humans have the ability of altering, both positively and negatively, our Earth's environment more than any of Earth's other inhabitants. Using the same innovative technologies already demonstrated in solving transportation, communication, crop yield, and energy exploration issues, we can also forestall much of the negative aspects of global warming and climate change by adapting to increasing global temperatures, rising sea levels, and extreme weather conditions. The solution to the perceived problems associated with global warming may not have to be the difficult solutions, but maybe the more manageable solutions.

We don't have to build on the beach and we can build better barriers to protect coastal cities from tidal surges. We can solve draughts with better water transportation solutions. If we can transport crude oil from the wells on the north slope of Alaska and the interior of Siberia across the frozen tundra, deserts, and forests to major transportation centers, why can't we do the same to transport water to solve drought in any part of the world? We can build our structures to resist environmental conditions of earthquakes, wind, and rain. We already have that capability as shown in the survivability of western world cities to earthquakes compared to survivability of earthquakes occurring in third world cities. The solutions are in fact economic ones. Until science can quantify the extent of global warming caused by burning fossil fuels, why should we deny cheap energy sources to the third world to improve their quality of life and simply adapt to climate change in the meantime?

We have to determine the sense of urgency to make realistic policy changes as we transition to renewable energy sources. As has been presented, the global temperature has risen less than 2 degrees C in the last two hundred years. If that rise continues at the same rate or possibly slightly higher rates as some models predict, can we "live" with that? Can we adapt to that? Do we have ten years or a hundred years? Can science tell us we are approaching environmental tipping points that require the "must act now" policies? Or should act now without knowing that tipping point or the economic consequences and unintended consequences of such action?

The question before us is how best to be good stewards of the Earth knowing that we can do nothing or sacrifice significantly to influence the environment in which we live. Each action we take has a cost. With global warming we become so fixated on reducing CO_2 omissions that we neglect our primary objective: to improve our quality of life as well as that of the environment. If we invest our children's economic future on the debt created by carbon taxes, economic stagnation, and an austere lifestyle only to gain insignificant reduction in global warming, what have we gained? I would suggest that we humans can adapt with fewer problems to small global temperature increases, rising sea levels, and extreme weather patterns, than we can to polluted waters and air that have direct and immediate impacts on the health of humans and other living things.

CONCLUSION/AUTHOR'S ANALYSIS

"Whenever a theory appears to you as the only possible one, take this as a sign that you have neither understood the theory nor the problem which it was intended to solve."
— Karl Popper, Philosopher and Professor

My interest was piqued in the topic of global warming many years ago when I saw a chart that encircled one of the rooms in the Natural History Museum in Washington D.C. that depicted the historical temperature of our planet Earth. It showed Earth's temperature as constantly changing and always warming and cooling never anything drastic. I thought this chart was in conflict with what some environmentalists, scientists, and the news media were portraying at that time, and up to the present. When I returned to that museum a few years ago, I noticed the chart was no longer on display, at least not where I thought I had seen it. The following NOAA chart constructed in 2008 depicts that same data that caught my attention those years ago:

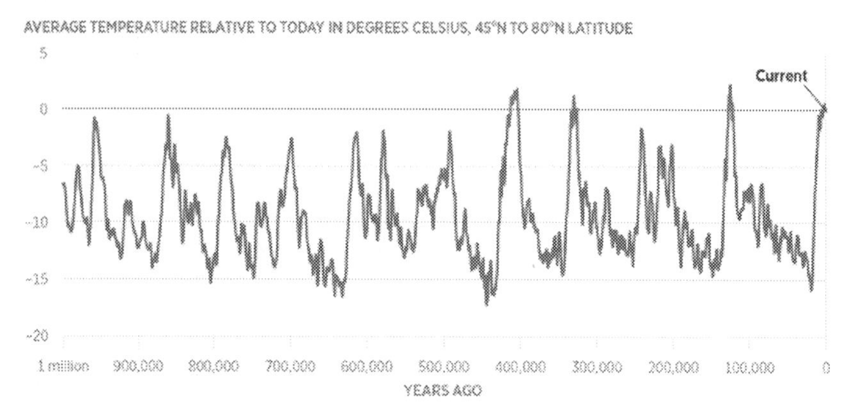

SOURCE: R. Bintanja and R.S.W. van de Wal, "Global 3Ma Temperature, Sea Level, and Ice Volume Reconstructions," National Oceanic and Atmospheric Administration, August 14, 2008, https://www.ncdc.noaa.gov/paleo/study/11933 (accessed April 5, 2016).

BG 3119 heritage.org

I have attempted in this book to present a broad view of our environment and the footprint we humans are leaving upon it, based on scientific evidence free to the extent possible from political perspective. I presented the issues and tried not to judge any presentation by its source but rather the substance of the presentation. To question the source rather than the substance would make the presentation political.

Our concerns with our environment in the last few decades have been, until just a few years ago, dominated by the anthropogenic impacts on global warming and its effect on climate change. These concerns, in my opinion, have overshadowed the concerns of the pollution of our lakes, rivers and atmosphere; how we dispose of our trash; the energy we use to heat and cool our homes, transport and prepare our food as it moves from the farm to our dining tables; and how we move about in our ever increasing mobility as citizens of the planet. It is more than sustaining, recycling, reusing, and conserving our resources; it is all of these and more. Our environment is enormously complex, and the science that drives it all, I believe, is far from understood.

Cleaning up our polluted lakes, rivers, and land mass should be the easy part, but on the global level it has taken a back seat to "cleaning" up the atmosphere of a colorless, tasteless, and odorless trace gas called CO_2; so essential to life on this planet. No other environmental concern has become as politically charged as has our concern with CO_2 and how to deal with it. The politics of CO_2 has manipulated the science as to how it is viewed from the warmist, the skeptic and the denier.

Scenes such as these of the Bagmati River in Nepal or the before and after picture of the Animas River in Colorado after a chemical waste spill from a mine in Colorado below should move us to action.

Instead our collective environmental concerns and resources are allocated more to the GHGs we humans emit into the atmosphere, while not taking into account the impact of the GHGs emitted naturally into the atmosphere as shown in this picture of GHG emissions from a volcanic emission.

CO_2 emissions have been made the target of the environmental movement as the cause of accelerated global warming and abnormal climate change. What does science tell us about CO_2 and its relationship to global warming and climate change? First we know that CO_2 is a trace gas with a current concentration of just over 400ppm, an increase from 250ppm in the middle of the eighteenth century at the start of the Industrial Revolution. Imagine, if you will, what a million marbles looks like. The figure below shows the relationship of the size of the Earth—a million would fill the volume of the sun. Now try to visualize those "Earths" being atmospheric molecules within which 400 CO_2 molecules are present.

Can you find them? And as explained in chapter 3 only 1.5 percent or four of those CO_2 molecules were emitted by human activity. Even scientists look at this and must ask themselves, "How much influence can such a small percentage emitted by humans of the total amount of that trace gas have on the atmosphere? What is the influence of those naturally-produced 396 molecules compared to the four produced by humans?" We do know that CO_2 and other greenhouse gases do trap heat as explained in chapter 3 and that can be scientifically shown. But if our hypothesis is increasing CO_2 will cause increased global temperatures, then we should be able to show significant global warming in the last two decades when CO_2 concentration increased by almost 20 percent. Our scientific temperature measurements, flawed as they may be, as shown in chapter 3, do not show a more significant global warming over that period, compared to the temperature trends previous to that. This lack of measureable warming caused the IPCC to walk back the warming forecast that their modeling had predicted (see footnote 4 below).

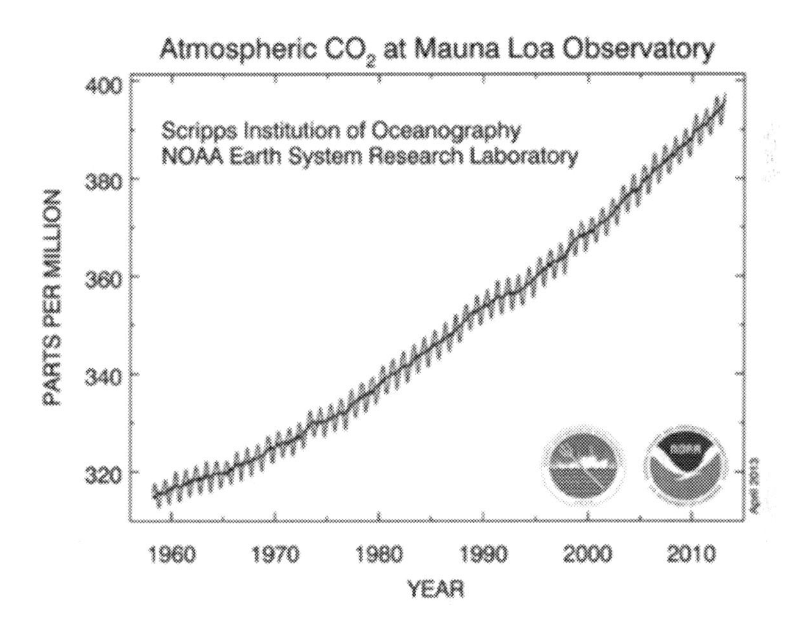

Graph Courtesy of NOAA

The scientific fact is that the science of climate and global warming is not well understood. Because CO_2 emissions have been increasing while temperatures have not, tells us for sure there are other forcings affecting the temperature control of the atmosphere. For sure the oceans are part of global temperature control, but that science is also not well known or understood. We know that the oceans hold a lot of heat, but that only the top meter or so, the **thermocline,** is influenced by atmospheric temperature changes, and the deep ocean heat is circulated by ocean currents such as El Niños.[1]

It is difficult not to see the big picture that the composition of our atmosphere is 78 percent nitrogen, 20.5 percent oxygen and all other gases 1.5 percent (chapter 3). So the remaining 1.5 percent contains all the greenhouse gases, plus some minute amounts of others. Of the total greenhouse gases, 95 percent is water vapor which is a constant over which humans have no control. Though the percentage varies slightly, our EPA supports the scientific research that about 3 percent of atmospheric CO_2 emissions are attributable to anthropogenic activity.[2] Therefore 97 percent of those CO_2 emissions occur naturally. Clearly we are talking about some very small, minute, amounts that humans contribute to global GHG emissions.

A point that I have not stressed enough is the benefit of increased CO_2 concentrations which add to the quality of human life and the health of the planet. It cannot be overstated that without CO_2, there can be no life on earth. The fact that CO_2 is added to commercial greenhouses[3] to aid in the growth of their plants speaks volumes. In his testimony on possible harmful effects of CO_2 before the U.S. Senate, Dr. William Happer, an American physicist, said CO_2 concentrations of 1000ppm would not be a "catastrophic" concentration of CO_2 in the atmosphere.[4]

CO_2 is essential for life and quite possibly more rather than less of it would be beneficial to life on Earth. We want to care for our environment, but it is not a given that the best way to help is to cut CO_2. A greener and warmer Earth is a healthier Earth and in chapter 3 we learned that vegetation growth on Earth is increasing as a suspected result of increased concentrations of CO_2. "We need to rein in exaggerations and start focusing on where we can do the most good" (Lomborg 2010). A greener and warmer Earth is a healthier Earth.

The Basic Science

Most who debate global warming will claim that science is on their side. To make that claim they must adhere to "good science." The scientific method tells us that science starts with an observation. Background research of the observation is then conducted, after which a hypothesis is constructed and then tested through experimentation. The data is then analyzed, a conclusion is drawn, and the results are shared. If our hypothesis is that increasing CO_2 causes global warming, but then our observation, testing, and measurements have not shown any significant warming for the past two decades while CO_2 emissions during that same period has continued to increase, then our hypothesis is wrong and must be modified.

Another concept in science made by a philosopher of science, Karl Popper in the mid twentieth century stated theories and hypotheses must also be falsifiable,[5] meaning capable of being tested, verified, or falsified by experiment or observation. He further stated theories which are not falsifiable are unscientific and therefore "pseudoscience." Simply put, we must be able to prove a theory false before we know if it could be true. For example horoscopes are not scientific because there is no way to disprove claims that are broad, vague and/or contradictory. Conversely Isaac Newton's claim that for every action there is an equal and opposite reaction is scientific because we can test that theory. If we make the claim that global warming leads to more precipitation and also less precipitation or it leads to more hurricanes or fewer hurricanes, or when it makes wet places drier and dry places wetter that is unscientific because if any and all climate conditions are being attributed to global warming/climate change than how can we design any observation or experiment falsifiable. A theory that explains everything in my mind, doesn't explain anything.

Clearly I think it can be shown that most of the recent increases in CO_2 can be attributed to human-caused emissions, but it is up to science to tell us how much impact this is having on our environment. Is there any experiment or observation that can distinguish between what is anthropogenic global warming/climate change and what are the natural causes of global warming/climate change? If the IPCC and warmist scientists want the public to take the theory of anthropogenic climate change seriously they need to become a lot more disciplined. Blaming the anthropogenic burning of fossil fuels for all things related to climate change and promoting a sense of urgency to quickly reduce CO_2 emissions to avoid climate disaster does not pass the test of the scientific method approach.

As has been shown, in the last one hundred fifty plus years CO_2 has increased by 45 percent yet the Earth has warmed only about 2 degrees F. Further it was shown that IPCC and NASA temperature modeling have overestimated temperature increase and especially in the last two decades causing both the IPCC and NASA to re evaluate their projections and modeling algorithms.[6] Again, if the observations don't support the hypothesis, the hypothesis is wrong. Obviously the models and hypotheses are flawed. Missing in the scientific research is the presence of an accurate accounting for forcings such as clouds, **ocean circulation**, volcanoes, and for sure solar activity. I believe the models that have been relied on to predict global warming have over simplified the impact of CO_2.

What Can We Do

We need to find smarter, cheaper, and more effective solutions to climate change. The fact remains that global warming is real, important and to an unknown extent man made, and to an unknown extent made by natural variations. We are, after all still emerging from the last ice age. It is self-defeating to shortcut the global warming debate with an end justifies the means approach that is morally and scientifically indefensible as well. There are those who believe that burning fossil fuels and cutting down our forests are the two driving forces of climate change. Those same people believe if we change the type of light bulb we use, or build wind turbines, or drive electric cars, we can reverse the catastrophic change in the climate they predict. If we believe this, then we believe humans are causing dangerous global warming. This idea that man is causing dangerous global warming is accepted today by "more than 190 heads of state, the United Nations, most scientific organizations, most Fortune 500 companies, the news media, and most universities.[7] The world is spending billions of dollars each year to try to stop the planet from warming. It would seem we need to ask three questions.

First, we have been able to measure that the Earth and its atmosphere has warmed in the last century and a half, but is that 2.5 degrees F of warming significant? More importantly is that temperature rise "dangerous"?

Second, if we are going to base public policy on the supposition that global warming is harmful and caused at least in part by humans, should we not quantify the amount of warming that is anthropogenic so we can know how best to deal with mitigating those harmful human contributions?

Third, what drives the need for immediate action? It seems like this is more a desire to capitalize on the emotion of the moment to do "at least something," no matter the consequences.

In 1970 we were told by Harvard biologist, George Wald, riding a wave of a popular environmental panic that "Civilization will end within fifteen or thirty years unless immediate action is taken…"[8] In 1977 Barack Obama's future science "czar," John Holdren, co-authored a book with Paul Ehrlich that global warming and overpopulation could lead to the deaths of one billion people through starvation by 2020.[9] (I think it is worth noting that the worst famines of the twentieth century were caused not by environmental disasters but by collectivists trying to control human nature.) In 2006 in his Oscar-winning documentary Al Gore warned that sea levels would rise by twenty feet "in the near future." The producers of that documentary offered chilling depictions of coastal cities under water. At a 2005 conference of climate scientists (and politicians) in London warned the world has as little as ten years before it would reach "the point of no return on global warming."[10]

Civilization moves on in spite of the all alarms, while the earth's population continues to grow with lower hunger issues than in the 1970s, as poverty continues to decline and South Beach is still with us. The problem for those who would require urgent change to save humanity is that humanity plods onward, living longer, safer, richer lives.

As the comedian George Carlin referred to the environment in one of his monologues, "It's not the Earth that's in trouble it's the [expletive deleted] people." Well he was so right. No matter what we humans do to our planet, it will continue on until cosmic forces destroy it—not we humans. To be sure, human survival on this planet depends on ability to preserve and care for our environment that sustains us, and to do that we must wisely maintain good stewardship of it.

In our concern for our environment we have overestimated the impact of CO_2 on atmospheric and global surface temperature heating, and underestimated the impact of the sun. We do not understand the role of clouds and the oceans in the warming and cooling process and certainly we do not understand the extent of the forcings that influence global warming or cooling. Climate science has a long way to go before it can be used to accurately advise on public policy. Until that time, we have much work to do in cleaning up what we know is harmful to our environment. We can clean up our waterways and invest in that cleanup throughout the world. We can invest in clean air, not by overburdening the planet with restriction of CO_2 emissions but the use of safe nuclear power and the renewable energy technologies of wind and solar when they become an economically competitive energy sources able to sustain itself free from taxpayer subsidies. Until then there is no immediacy to disrupt our lifestyles and economic prosperity by banning the use of the cheap energy we get from of fossil fuels. We can, however, burn them more efficiently through increased use of clean coal plants, continue the transition to the use of natural gas, and the use of more efficient internal combustion engines (with which the auto industry has done really well). These efforts will emit less particulate pollution in our atmosphere and promote cleaner and healthier atmosphere. The effort must be global. We cannot deny cheap energy sources to the developing world as economic prosperity is required to clean up and care for the environment of our planet. We must make energy use more efficient rather than restrictive. We must emphasize reuse rather than recycle in geographic areas that make it more economical, recycle when it is more efficient than incinerating, and landfill as a last resort.

We can enhance the sustainability of our environment by using the resources of our Earth wisely, not driven by knee-jerk restrictions or dogma but by sustaining the good qualities of supportive and renewable resources of our Earth that we would want to pass on to our children.

Then when your scoutmaster calls out, "why did you cut down that tree?" you'll have a good answer.

GLOSSARY

Acid Rain—Acidic rainfall from waste gases of sulfur and nitrogen oxides, which combine with atmospheric water to form acids; caused primarily from the industrial burning of coal and other fossil fuels that causes atmospheric pollution and environmental harm, typically to forests and lakes.

Active Environmentalist—One who participates in sound environmental behavior. (Author's definition.)

Adaptation—Altering human behavior to cope with change, and taking measures in construction/habitation techniques to cope with climate impacts. eg building reservoirs as a defense against drought and dykes against sea level rise or sowing heat-resistant crops.

Air Capture—A technology associated with the capture of CO_2 from atmospheric air.

Alarmist—The concerned environmentalists whose view is global warming and climate change is a real and immediate threat and for the most part believe human causes are driving these environmental changes.

Alliance for Climate Protection—Al Gore founder and Chairman.

Albedo—The proportion of solar energy reflected back into space. The presence of clouds increases the Earth's albedo.

ANNEX I & ANNEX II—A division between rich and developing countries enshrined in the convention of the UN's climate forum in 1992, which places emission reduction targets on Annex I countries, responsible for historic emissions.

Anthropogenic—Originating from human activity.

Anthropocene—The age of humankind.

AOSIS—The Alliance of Small Island States (AOSIS), highly vulnerable to sea-level rise.

ARs 1-5—The progression of Intergovernmental Panel on Climate Change (IPCC) assessment reports on global warming and its impacts.

Artificial Trees—Used in applications to save natural trees.

Atmosphere Heat Transport—The movement of heat in the atmosphere through conduction, radiation, and convection.

Biochar—Charcoal produced from plant matter and stored in the soil as a means of removing carbon dioxide from the atmosphere.

Biodegradable—A substance or material capable of being decomposed by bacteria or other living organisms. Material that comes from either a plant or animal.

Biofuel—A fuel derived directly from living matter.

Biogenic—Resulting from biological activity or from living things.

Biosphere—The regions of the surface, atmosphere, and hydrosphere of the earth occupied by living organisms.

Bio Mass—Plant material and animal waste, the oldest source of renewable energy, used since our ancestors learned the secret of fire.

Black Carbon—Soot.

Borehole—Used to bring to the surface an ice core (cylinder) from ice sheets and glaciers. They are essentially frozen time capsules that allow scientists to reconstruct climate far into the past.

Butterfly Effect— The phenomenon whereby a minute localized change in a complex system can have large effects elsewhere.

Cap and Trade—Buying and selling unused carbon emissions quotas under a cap, or ceiling, imposed on a country, region or industrial sector.

Cape Grim—Cape Grim, on Tasmania's west coast, is one of the three premier Baseline Air Pollution Stations in the World Meteorological Organization-Global Atmosphere Watch (WMO-GAW) network. Baseline stations are defined by the WMO to meet a specific set of criteria for measuring greenhouse and ozone depleting gases and aerosols in clean air environments.

Carbon Credits—A carbon credit is a generic term for any tradeable certificate or permit representing the right to emit one ton of carbon dioxide or the equivalent amount of a different greenhouse gas.

Carbon Capture and Storage (CCS)—A fledgling technology to siphon and store carbon waste (CO_2) from fossil fuel-burning power plants transporting it to a storage site, and depositing it where it will not enter the atmosphere, normally an underground geological formation.

Carbon Dioxide Removal (CDR)—Refers to a number of technologies which reduce the levels of carbon dioxide in the atmosphere.

Carbon Leakage—Leakage that occurs when there is an increase in carbon dioxide emissions in one country as a result of an emissions reduction by a second country with a strict climate policy.

Carbon Tax—A tax levied on the carbon content of fuels. It is a form of carbon pricing. Carbon is present in every hydrocarbon fuel and converted to carbon dioxide and other products when combusted.

COP—Conference of Parties, the UN climate forum's decision-making body, which meets once a year. The November 30-December 11, 2018 meeting in Paris was the twenty-first meeting, hence COP 21.

Carbon Dioxide Removal (CDR)—Refers to a number of technologies of which the objective is the large-scale removal of carbon dioxide from the atmosphere.

Clean Burning—Leaving little contamination while consuming a fuel; The "cleanest burning" is that which produces only CO_2 and H_2O from the combustion of hydrocarbons.

Climate Change—Any substantial change in Earth's climate that lasts for an extended period of time.

Climate Deniers—Those who dismiss human-caused climate change and global warming as unwarranted, and unsupportable.

Climate Engineering—The deliberate and large-scale intervention in the Earth's climate system by human means.

Climate Feedbacks—An interaction between various processes of climate system is called climate feedback, where in the result of one process triggers changes in the another process that in turn will influence the initial one. Positive feedbacks amplify effects and negative feedbacks diminish effects.

Climate Proxies—These are preserved physical characteristics of the past preserved in ice cores, tree rings, boreholes, lake and ocean sediments that can be used as stand ins to make direct measurements which enable scientists to reconstruct the climate conditions that prevailed during the Earth's history.

Climate Sensitivity—defined as the amount of global average surface warming following a doubling of carbon dioxide concentrations.

Climate Skeptic—Generally the label applied to those who are either skeptical that global warming is occurring and/or skeptical that warming is caused by anthropological contribution.

Climate Warmist—Generally the label applied to those who are convinced global warming is reality and for the most part caused by anthropological contribution.

Community Earth System Model—developed by the scientists at the North Carolina University for Atmospheric Research and considered to be one of the most sophisticated models of global climate.

Composting—setting aside vegetable matter or manure from ordinary human waste to decay and then add to soil as a fertilizer.

Copenhagen Summit—A United Nations Climate Change Conference held in December 2009 where the framework for climate change mitigation beyond 2012 was to be agreed there.

Coupled Modeling Intercomparison Project (CMIP)—Formulated by 20 modeling groups from around the world who agreed to promote a new set of climate modeling experiments to assess why similarly forced models produce a range of responses.

Cryogenian—A geological period that last from 720 to 635 million years ago during which the Earth experienced it s two greatest ice ages; the Sturtian and Marinoan.

Cryosphere—Those portions of Earth's surface where water is in solid form, including sea ice, lake ice, river ice, snow cover, glaciers, ice caps and ice sheets, and frozen ground (which includes permafrost). It is an integral part of the global climate system.

Demographics—The statistical data relating to any given population and the particular groups within it.

Denier—Those who do not believe global warming is a reality and/or that human intervention has much if any impact on climate change.

Direct Air Capture (DAC)—The capture of CO_2 from the atmosphere.

Earth Stewardship—The belief that humans are responsible for the world, and should take care of it.

El Niño—A band of irregular and strange warm ocean water temperatures that occasionally develops off the western coast of South America which can cause climatic changes across the Pacific Ocean. These temperatures oscillate from causes unknown.

ENSO (El Niño—Southern Oscillation)—Irregularly periodical variation in winds and sea surface temperatures over the tropical eastern Pacific Ocean, affecting much of the tropics and subtropics. The warming phase is known as El Niño and the cooling phase as La Niña.

Environmental Activist—One who participates in environmental issues through political campaigns and social movements. (Author's definition)

Entropy—A thermodynamic quantity representing the unavailability of a system's thermal energy for conversion into mechanical work, often interpreted as the degree of disorder or randomness in the system.

E-waste—Discarded electronic products such as computers, televisions, VCRs, copiers, etc. E-waste is an environmental concern because it contains toxic substances such as lead and mercury.

Feedbacks—Climate reactions which alter the response of the system to changes in external forcings. Positive feedbacks increase the response of the climate system to an initial forcing, while negative feedbacks reduce the response of the climate system to an initial forcing. Uncertainty over the effect of feedbacks is a major reason why different climate models project different magnitudes of warming for a given forcing scenario.

Feedback Loop—A biological occurrence wherein the output of a system amplifies the system (positive **feedback**) or inhibits the system (negative **feedback**). **Feedback loops** are important because they allow living organisms to maintain homeostasis (stable equilibrium).

Feedback Mechanisms—Clouds, evaporation, snow, and ice interaction with each other.

Forcings (negative and positive)—Measure of the influence a factor has in altering the balance of incoming and outgoing energy.

Forcing Agents—Factors that affect the Earth's climate. These "forcings" drive or "force" the climate system to change.

Fracking—is the process of drilling down into the earth and a mixture of water, sand, and chemicals are injected into the rock at high pressure which allows the gas to flow out to the head of the well. This process has also been used as a method of nuclear waste disposal.

Fresh Water—Natural occurring water except seawater, includes water in ice sheets, ice caps, glaciers, icebergs, bogs, ponds, lakes, rivers, streams, and even underground water called groundwater.

Geochemistry—The study of the chemical composition of the earth and its rocks and minerals.

Geochronology—The branch of geology concerned with the dating of rock formations and geological events.

Geo Engineering—The deliberate, large-scale manipulation of the Earth's climate system to counteract climate change Climate engineering refers to the deliberate large-scale manipulation of the Earth's environment to counter the effects of climate change through large-scale strategies to reduce global warming by removing carbon dioxide from the atmosphere or reducing solar input to Earth.

GISS—Goddard Institute for Space Studies. A NASA laboratory in the Earth Sciences Division to do basic research in space sciences.

Glaciology—The study of the internal dynamics and effects of glaciers.

Global Climate Modeling—The use of mathematical equations to describe the behavior of the factors of the Earth system that impact climate; some of which for example, are dynamics of the atmosphere, oceans, land surface, living things, and ice, plus energy from the Sun.

Global Circulation Model (GCM)—A modeling tool available for simulating the response of the global climate system to increasing greenhouse gas concentrations.

Global environmental governance—The sum of organizations, policy instruments, financing mechanisms, rules, procedures and norms that regulate the processes of global environmental protection.

Global Warming—The observed century-scale rise in the average temperature of the Earth's climate system and its related effects

Global Warming Alarmists—The view taken by those who believe global warming is happening, is caused primarily by human involvement and with real potential of irreversibly changing our environment and climate.

Global Warming Deniers—The view that global warming and/or its effects are not really happening.

Global Warming Skeptics—The view taken that the human causes of global warming are not known, and/or not convinced human-caused global warming is even happening.

Globalization—Refers to the process of the intensification of economic, political, social and cultural relations across international boundaries. It is principally aimed at the homogenization of political and socio-economic theory across the globe. It is a process by which regional economies, societies, and cultures have become integrated through a global network of communication, transportation, and trade.

Green Climate Fund (GCF)—A fund designed to channel money in climate aid into poor countries.

Greenhouse Effect—The long wave terrestrial radiation emitted by the warm surface of the Earth is partially absorbed and then re-emitted by a number of trace gases in the cooler atmosphere above. On average, the outgoing long wave radiation balances the incoming solar radiation both the atmosphere and the surface will be warmer than they would be without greenhouse gases.

Greenhouse gases (GHGs)—The gases that have the capacity to retain solar heat in Earth's atmosphere. They are Water Vapor, Carbon Dioxide (CO_2), Methane (CH_4), Ozone (O_2), Nitrous Oxide (N_2O), Chlorofluorocarbon-12 (CCl_2F_2), Hydroflorocarbon-23 (CHF_3), Sulfur Hexafluoride (SF_6), and Nitrogen Triflouride(NF_3). The first five are the main GHGs considered for their global warming potential.

Green Up—The concept of changing behaviors to more environmentally friendly ones.

Heartland Institute—An American conservative and libertarian public policy think tank founded in 1984.

Heat Islands—An urban area or metropolitan area that is significantly warmer than its surrounding rural areas due to human activities such as urbanization e.g. concrete structures, roads, parking lots etc.

Homogenization—A leveling of the playing field for data used in modeling justified on the basis that corrections must be made and non climatic variable changes removed when comparing one weather station to another.

Hydrofracture—A process by which water is injected into low yielding water wells at a high pressure and volume. The theory is to open up and clean out the existing fractures found in the rock structure of a well.

Hypothesis—A supposition or proposed explanation made on the basis of limited evidence as a starting point for further investigation.

Ice Age—Also referred to as glacial periods. A period of long term reduction in the temperature of the Earth's surface and atmosphere, resulting in the presence or expansion of continental and polar ice sheets and alpine glaciers.

Ice Cores—Cylinders of ice drilled out of an ice sheet or glacier. Most ice core records come from Antarctica and Greenland, and the longest ice cores extend to 3km in depth. The oldest continuous ice core records to date extend 123,000 years in Greenland and 800,000 years in Antarctica.

Incontrovertible—Not able to be denied or disputed.

Industrial Revolution—The revolution marked a major turning point in history beginning in the late eighteenth and early nineteenth centuries, characterized by the rapid development of industry. This brought about by the introduction of machinery that required far more combustion of fossil fuels to power its rapid growth in the production of manufactured goods and modes of transportation in the transition from production by hand to that of machines.

Inert Waste—Waste that will not decompose, because it is not biologically reactive. Examples are sand and concrete.

Interglacial Period—A geological interval of warmer global average temperature lasting thousands of years that separates consecutive glacial periods within an ice age.

IPCC—Intergovernmental Panel on Climate Change.

Isotope—The radioactive form of an element, each of two or more forms of the same element that contain equal numbers of protons but different numbers of neutrons in their nuclei, and hence differ in relative atomic mass but not in chemical properties

James Hansen—An American adjunct professor directing the Program on Climate Science, Awareness and Solutions of the Earth Institute at Columbia University who directs NASA's Goddard Institute for Space Studies

Kyoto Protocol (KP)—Through the United Nations Framework Convention on Climate Change (UNFCCC), this is an international treaty that set binding obligations on industrialized countries to reduce emissions of greenhouse gases. The United States signed but did not ratify the Protocol and Canada withdrew from it in 2011. This 1997 accord on carbon emissions is to be replaced by the 2015 Paris agreement.

La Niña—This is a band of irregular and strange cool ocean water temperatures that occasionally develops in the western Pacific which can cause climatic changes across the Pacific Ocean. The opposite of El Niño. As with El Niño these temperatures oscillate from causes unknown.

Leadership in Environmental and Engineering Design (LEED)—The standardization agency for environmental building codes.

LDCs—Least Developed Countries

Little Ice Age—A period of cooling, but not a true ice age, that occurred after the Medieval Warm Period from around 1300 to 1870 Marcott-Shakun Dating Service A temperature data analysis that omitted some data showing warming was excluded from the 1940 report which lent support to the "hockey stick" graph.

Margin of Error—An amount (usually small) that is allowed for in case of miscalculation or change of circumstances.

Marine Cloud Brightening—A geoengineering concept that would seed marine stratocumulus clouds with substantial concentrations of seawater particles to enhance cloud albedo and longevity, thereby producing a cooling effect.

Medieval Warm Period (MWP)—A period of unusually warm weather that began around 1000 AD and persisted until a cold period known as the "Little Ice Age" took hold in the 14th century. Warmer climate brought a remarkable flowering of prosperity, knowledge, and art to Europe during the High Middle Ages, and Norse exploration of the northern hemisphere.

Meteorology—The branch of science concerned with the processes and phenomena of the atmosphere, especially as a means of forecasting the weather, the climate and weather of a region.

Methanosarcina—Single-celled organisms that are known as anaerobic methanogens which produce methane using all three metabolic pathways for methanogenesis.

Mitigation—Measures to reduce or slow the emissions of greenhouse gases that cause global warming.

Mutagen—A physical or chemical agent that changes the genetic material, usually DNA, of an organism and thus increases the frequency of mutations above the natural background level.

National Academy of Science (NAS) and National Resource Council (NRC)—A private, non-profit society of distinguished scholars. Established by an Act of Congress, signed by President Abraham Lincoln in 1863, the NAS is charged with providing independent, objective advice to the nation on matters related to science and technology.

NASA—National Aeronautics and Space Administration. An independent agency of the executive branch of the federal government. It is responsible for both the Johnson Space Center in Houston and the Goddard Institute for Space Studies (GISS)

Net energy—A concept used in energy economics that refers to the difference between the energy expended to harvest an energy source and the amount of energy gained from that harvest.

NOAA—National Oceanic and Atmospheric Administration. The organization of scientists that study the skies and oceans.

Nongovernmental International Panel on Climate Change—A nongovernmental International panel of scientists and scholars that come together to try to understand the causes and consequences of climate change. No attachment to any government funding, making it independent from any political pressure or influences or predisposed to any political agenda or policies.

Nuclear Winter—The term given to the theory that post nuclear detonations would cause massive amounts of smoke and debris into the atmosphere that would cool the planet.

Null Hypothesis—Refers to a general or default position: that there is no relationship between two measured phenomena

Ocean Acidification—Often referred to as the "other CO_2 challenge" is the name given to the ongoing decrease in the pH of the Earth's oceans, caused by the absorption of carbon dioxide (CO_2) from the atmosphere. Ocean acidification, which is driven by increased levels of atmospheric carbon dioxide, has been regarded by many climate scientists and geoengineers as the "equally evil twin" of global climate change, arguably affecting fish coral reefs and other sea flora to varying degrees. A misleading term as the oceans average a pH of around 8 are alkaline not acidic. As the oceans warm and cool CO_2 concentrations increase and decrease causing constant changes in pH levels.

Ocean Anoxia—An event that occurs when oceans become completely depleted of oxygen.

Ocean Circulation—Movement of sea water generated by a number of forces acting upon the water, including wind, the Coriolis effect, breaking waves, and temperature and salinity differences.

Ocean Fertilization—To increase marine food production by the purposeful addition of nutrients to the oceans' surface which would remove carbon dioxide from the atmosphere by, then absorbed by plankton which would sink, taking the absorbed CO_2 with it.

Ocean Heat Transport—The movement of ocean water from the equator to the poles driven by the wind and generally confined to the thermocline.

OMEGA—Offshore Membrane Enclosures for Growing Algae. A NASA pilot program for CO_2 reduction that mimics the natural production of algae blooms in a controlled ocean environment.

Opacity—The condition of lacking transparency or translucence.

Organization for Economic Co-operation and Development (OECD)—An intergovernmental economic organization with 35 member countries, founded in 1961 to stimulate economic progress and world trade.

Oxygen isotopes 16 and 18—Oxygen is made up of two isotopes: oxygen 16 and oxygen 18. The relative amounts of these two isotopes present in any sample of water, ice, rock, plant, animal, etc. is a function of climate/environment.

Pacific Decadal Oscillation (PDO)—A pattern of large scale Pacific climate variability similar to ENSO (El Niño—Southern Oscillation) in character, but which varies over a much longer time scale. The **PDO** can remain in the same phase for 20 to 30 years, while ENSO cycles typically only last 6 to 18 months.

Paleoclimatology—The study of changes in climate taken on the scale of the entire history of Earth.

Paleontology—The branch of science concerned with fossil animals and plants.

Paleoecology—The ecology of fossil animals and plants.

Permian Triassic Extinction Event—Known as the "Great Dying." it occurred 250 million years ago. It is the Earth's most severe known extinction event, with up to 96 percent of all marine species and 70 percent of terrestrial vertebrate species becoming extinct. It is the only known mass extinction of insects.

Photovoltaic—Relating to the production of electric current at the junction of two substances exposed to light.

Pleistocene Ice Age—The Pleistocene Epoch is typically defined as the time period that began about 2.6 million years ago and lasted until about 11,700 years ago. The most recent Ice Age occurred then, as glaciers covered huge parts of the planet Earth.

PPM—Parts per million.

Polyethylene Terephthalate or PET—A globally recognized safe, non toxic, and recyclable plastic packaging material for the storage of food and beverages.

Proxies—Provide a means for scientists to determine climatic patterns before record-keeping began. Examples of proxies include tree rings, boreholes, sea and lake sediments, and corals.)

Pyrolysis—The subjection of organic compounds to very high temperatures and the resulting decomposition.

R3—Reduce, reuse and recycle; the three essential components of environmentally-responsible consumer behavior.

Radiative Forcing—The difference between the energy absorbed by the Earth and that emitted by it back into space. A positive forcing (more incoming energy) warms the system, while negative forcing (more outgoing energy) cools it. It is quantified in units of watts (rate of energy conversion or transfer with respect to time) per square meter of the Earth's surface.

Radio Carbon Dating— A method for determining the age of an object containing organic material by using the properties of radiocarbon, a radioactive isotope of carbon.

RCP—Representative Concentration Pathways, A modeling input that allows for varying modeling inputs to illustrate uncertain outcomes; a projection of varying impacts.

Recycle—Converting (through various forms of treatment processes) waste into reusable material.

Reforestation—The natural or intentional replanting of existing forests and woodlands that have been depleted, usually through deforestation.

Renewable Energy—Generally considered most environmentally friendly and practical forms of energy, includes Wind, Solar, Biomass, and Hydro.

Reuse—To use again a resource either for its original function or creatively for a different function.

RSS—Remote sensing systems (satellites).

Sedimetology—The science of studying the history of the planet through its sedimentary layers.

Sequestration—To seize or capture.

Skeptic—Those concerned environmentalists who are not sure of the causes of global warming or any perceived climate change.

Sky Whitening—The fading of blue skies to white if aerosols are injected into the atmosphere.

Soil management—Concerns all operations, practices, and treatments used to protect soil and enhance its performance.

Solar Forcing—The difference between insolation absorbed by the Earth and energy radiated back to space.

Solar Radiation Management (SRM)—The introduction of means to prevent solar energy from reaching the planet.

Special Report Emissions Scenarios—A report by the IPCC used to make projections of future climate change. These reports have been replaced by Representative Concentration Pathways (RCPs)

Spectroscopy—The study of spectra to determine the chemical composition of substances and the physical properties of molecules, ions, and atoms.

Stratigraphic—The study of sedimentary and layered rocks

Tectonics—Large-scale processes affecting the structure of the earth's crust.

Temperature Proxies—Alkenone, planktonic foraminifera Mg/Ca 23, fossil pollen, ice-core stable isotopes, tree ring reconstructions.

Tetrogen—Any agent that can disturb the development of an embryo or fetus. They may cause birth defects or terminate pregnancy outright.

Thermocline—A steep temperature gradient in a body of water such as a lake, or ocean marked by a layer above and below which the water is at different temperatures which tends to prevent mixing between the surface waters and those beneath the thermocline

Tipping Point—Feedback loops, where climate change feeds back on itself and causes rapidly accelerating, catastrophic, and implying irreversible consequences.

Total Solar Irradiance (TSI)—The rate at which energy from the Sun reaches the top of Earth's atmosphere is called "total solar irradiance" (or **TSI**). **TSI** fluctuates slightly from day to day and week to week.

Twomey Effect—Describes how cloud condensation nuclei (CCN), possibly from anthropogenic pollution, may increase the amount of solar radiation reflected by clouds. This effect is indirect. This increases the cloud albedo as clouds appear whiter and larger, leading to a cooling. An example: we observe trails of white clouds from ships crossing the oceans due to this effect.

United Nations Framework Convention on Climate (UNFCCC)—The UN global treaty with the aim of curbing global warming.

U.S. Green Building Council (USGBC)—In conjunction with LEEDs, intended to provide building owners and operators a concise framework for identifying and implementing practical and measurable green building design, construction, operations and maintenance solutions.

Urban Heat Island (UHI) Effect—The tendency for cities to be warmer because they retain the daytime solar radiation in the concrete, brick, and asphalt infrastructure and then release the heat that night as infrared, thus warming the air that would normally cool.

Warmist—A term used to signify those who believe global warming is occurring because of by human causes.

ENDNOTES

Introduction

1. www.google.com/search?q=images+of+al+gore+inconvenient+truth&source=lnms&tbm=isch&sa=X&ved=0ahUKEwi73NjkrIjiAhXopVkKHcJlBxgQ_AUIDygC&biw=1280&bih=921#imgrc=biZy6WEj7_2TEM:

Chapter 1

1. www.scholastic.com/teachers/articles/teaching-content/history-native-americans/
2. www.perc.org/1996/07/01/conservation-native-american-style/
3. Breaking the Environmental Policy Gridlock, Terry L. Anderson, 1997
4. www.traditionalanimalfoods.org/mammals/hoofed/page.aspx
5. Breaking the Environmental Policy Gridlock, Terry L. Anderson, 1997

Chapter 2

1. www.ossfoundation.us/projects/environment/global-warming/atmospheric-compositionl
2. www. socratic.org/questions/how-has-the-composition-of-the-earth-s-atmosphere-changed-over-time /
3. www.ncdc.noaa.gov/monitoring-references/faq/greenhouse-gases.php
4. www.weather.gov/jetstream/energy
5. www.butane.chem.uiuc.edu/pshapley/GenChem1/L15/web-L15.pdf
6. www.butane.chem.uiuc.edu/pshapley/GenChem1/L15/web-L15.pdf

7. www.wattsupwiththat.com/2014/07/29/epa-document-supports-3-of-atmospheric-carbon-dioxide-is-attributable-to-human-sources/
8. www.theguardian.com/environment/2012/jan/16/greenhouse-gases-remain-air
9. www.ib.bioninja.com.au/standard-level/topic-4-ecology/44-climate-change/greenhouse-gases.html
10. www.earthobservatory.nasa.gov/features/Arrhenius
11. www.smithsonianmag.com/travel/giant-hole-ground-has-been-fire-more-40-years-180951247
12. www.chinapower.csis.org/china-greenhouse-gas-emissions/
13. www.epa.gov/ghgemissions/overview-greenhouse-gases
14. www.epa.gov/ghgemissions/overview-greenhouse-gases
15. www.americanthinker.com/articles/2010/02/the_hidden_flaw_in_greenhouse.html
16. www.climate.ncsu.edu/edu/Milankovitch
17. www.climatebits.umd.edu/info/SolarRadiationInfo.html
18. www.scied.ucar.edu/molecular-vibration-modes
19. www.history.aip.org/climate/co2.htm
20. www.esrl.noaa.gov/gmd/outreach/info_activities/pdfs/CTA_carbon_on_the_move.pdf
21. www.lasp.colorado.edu/home/wp-content/uploads/2011/08/What-is-the-Carbon-Cycle.pdf
22. www.sciencing.com/photosynthesis-equation-6962557.html
23. www.sciencing.com/photosynthesis-equation-6962557.html
24. www.science.howstuffworks.com/environmental/earth/geophysics/earth3.htm
25. www.earthobservatory.nasa.gov/features/Water/page2.php
26. www.wattsupwiththat.com/2017/12/19/new-svensmark-paper-the-missing-link-between-cosmic-rays-clouds-and-climate-on-earth/

Chapter 3

1. www.livescience.com/40311-pleistocene-epoch.html
2. www.wattsupwiththat.com/2013/11/17/climate-and-human-civilization-over-the-last-18000-years/

3. www.wattsupwiththat.com/2013/11/17/climate-and-human-civilization-over-the-last-18000-years/
4. www.geocraft.com/WVFossils/GlobWarmTest/A6c.html
5. www.geocraft.com/WVFossils/Carboniferous_climate.html
6. www.globalclimate.ucr.edu/resources.html
7. www.learner.org/courses/envsci/unit/text.php?unit=12&secNum=3
8. www.history.aip.org/climate/co2.htm
9. www.learner.org/courses/envsci/unit/pdfs/unit12.pdf
10. www.randombio.com/co2.html
11. www.randombio.com/co2.html
12. www.scientificamerican.com/article/how-are-past-temperatures/
13. www.scotese.com/climate.htm
14. www.ncdc.noaa.gov/global-warming/temperature-change
15. www.joannenova.com.au/2009/12/carbon-rises-800-years-after-temperatures/
16. www.joannenova.com.au/2009/12/carbon-rises-800-years-after-temperatures/
17. www.joannenova.com.au/2009/12/carbon-rises-800-years-after-temperatures/
18. www.greenfacts.org/en/climate-change-ar5-science-basis/figtableboxes/figure-ts14.htm
19. www.huffingtonpost.com/2013/09/19/ipcc-climate-report_n_3957766.html
20. www.nasa.gov/press-release/nasa-noaa-analyses-reveal-record-shattering-global-warm-temperatures-in-2015
21. www.theatlantic.com/technology/archive/2013/05/the-hockey-stick-the-most-controversial-chart-in-science-explained/275753/
22. www.washingtontimes.com/news/2003/aug/25/20030825-090130-5881r/
23. www.researchgate.net/publication/237416644_The_Hockey_Stick_Debate_And_Due_Diligence_Lessons_in_Disclosure
24. www.epw.senate.gov/public/index.cfm/press-releases-all?ID=75771A77-802A-23AD-4F1E-F2E1E69F0BCF
25. www.cienceandpublicpolicy.org/wp-content/uploads/2010/07/ad_hoc_report.pdf

26. www.realclimate.org/index.php/archives/2012/07/tree-rings-and-climate-some-recent-developments/
27. www.theatlantic.com/technology/archive/2013/05/the-hockey-stick-the-most-controversial-chart-in-science-explained/275753/
28. www.climatecentral.org/gallery/maps/the-tale-of-two-marches
29. www.washingtonpost.com/politics/for-president-obama-a-renewed-focus-on-climate/2014/05/04/6b81412c-d144-11e3-9e25-188ebe1fa93b_story.html?utm_term=.65c4d09d1061
30. www.forbes.com/sites/jamestaylor/2014/10/09/noaa-destroys-global-warming-link-to-extreme-weather/#6b5c643350c8
31. www.oomg.meas.ncsu.edu/wp-content/uploads/2016/12/He_et_al-JAMES2015.pdf
32. www.judithcurry.com/2013/10/01/ipcc-solar-variations-dont-matter/
33. www.sistertoldjah.com/archives/2007/08/09/nasa-corrects-climate-figures-warmest-year-on-record-is-1934/
34. www.desmogblog.com/roy-spencer
35. www.history.aip.org/climate/solar.htm

Chapter 4

1. www.theglobalist.com/the-history-of-energy-since-10000-b-c/
2. www.dispatch.com/news/20180611/study-finds-no-evidence-of-groundwater-contamination-from-fracking

Chapter 5

1. www.un.org/documents/ga/res/43/a43r053.htm
2. www.un.org/documents/ga/res/43/a43r053.htm
3. www.climatechangereconsidered.org/about-the-nipcc/
4. www.climatechangereconsidered.org
5. www.economist.com/node/21555894/all-comments?page=4
6. www.ipcc.ch/ipccreports/far/wg_I/ipcc_far_wg_I_full_report.pdf
7. www.c2es.org/content/ipcc-fifth-assessment-report/

8. www.forbes.com/sites/realspin/2014/03/31/the-ipccs-latest-report-deliberately-excludes-and-misrepresents-important-climate-science/#2521ad00428e
9. www.heartland.org/_template-assets/documents/CCR/CCR-II/One-page-summary-of-CCR-II.pdf
10. www.heartland.org/_template-assets/documents/CCR/CCR-II/One-page-summary-of-CCR-II.pdf
11. www.heartland.org/publications-resources/publications/the-global-warming-crisis-is-over
12. www.climatechangereconsidered.org/about-the-nipcc/
13. www.csiro.au/greenhouse-gases
14. www.usnews.com/news/articles/2016-09-29/atmospheric-carbon-dioxide-levels-pass-400-ppm-tipping-point-maybe-for-good
15. www.drroyspencer.com/
16. www.climate.nasa.gov/scientific-consensus/
17. www.youtube.com/watch?v=NZq6zc0G018
18. www.ourchangingclimate.wordpress.com/2016/06/22/new-survey-of-climate-scientists-by-bray-and-von-storch-confirms-broad-consensus-on-human-causation
19. www.oism.org/pproject
20. www.petitionproject.org/
21. www.wnd.com/2014/04/u-n-climate-chief-admitted-political-agenda-of-science-report/
22. www.climatedepot.com/2013/08/21/un-scientists-who-have-turned-on-unipcc-man-made-climate-fears-a-climate-depot-flashback-report

Chapter 6

1. www.whitehouse.gov/about-the-white-house/the-executive-branch/
2. www.epa.gov/aboutepa/our-mission-and-what-we-do
3. www.arb.ca.gov/ba/omb/50thfinal/tsld017.htm
4. www.forbes.com/sites/markhendrickson/2013/03/14/the-epa-the-worst-of-many-rogue-federal-agencies/
5. www.epa.gov/aboutepa/ddt-ban-takes-effect

6. www.junkscience.com/1999/07/100-things-you-should-know-about-ddt/#ref7
7. www.timpanogos.wordpress.com/2015/08/26/audubon-christmas-bird-count-issue-eagles-did-not-prosper-during-the-time-of-ddt/
8. www.junkscience.com/1999/07/100-things-you-should-know-about-ddt/#ref7
9. www.junkscience.com/1999/07/100-things-you-should-know-about-ddt/#ref7
10. www.judithcurry.com/2015/05/06/quantifying-the-anthropogenic-contribution-to-atmospheric-co2/
11. www.epa.gov/newsreleases/epa-releases-2018-power-plant-emissions-demonstrating-continued-progress
12. www.dailycaller.com/2014/12/05/harvard-law-professor-epa-climate-rule-is-unconstitutional/
13. www.thenewamerican.com/usnews/politics/item/19130-report-details-close-relationship-between-epa-and-green-groups
14. www.dailycaller.com/2013/12/20/muzzled-epa-silenced-scientists-that-challenged-their-agenda/
15. www.dailycaller.com/2013/12/20/muzzled-epa-silenced-scientists-that-challenged-their-agenda
16. www.epa.gov/ttn/atw/utility/sec_111_respcom_oar-2002-0056-6206.pdf
17. www.fas.org/sgp/crs/misc/R42895.pdf
18. www.eia.gov/environment/emissions/carbon/

Chapter 7

1. www.reason.com/archives/2013/02/15/the-climate-and-energy-state-of-the-unio
2. www.resourcesmag.org/common-resources/understanding-the-mccain-lieberman-stewardship-act
3. www-pub.naz.edu/~rgecas7/problem.htm
4. www.epw.senate.gov/public/_cache/files/bc209836-a786-4792-bed0-cfcdef0fe649/final-epw-white-paper-lessons-from-kyoto-4.21.2016.pdf

5. www.govtrack.us/congress/bills/108/s139
6. www.eesi.org/papers/view/fact-sheet-carbon-pricing-around-the-world?/fact-sheet-carbon-pricing-around-world-17-oct-2012
7. www.theatlantic.com/business/archive/2012/06/26-trillion-pounds-of-garbage-where-does-the-worlds-trash-go/258234/
8. www.theatlantic.com/business/archive/2012/06/26-trillion-pounds-of-garbage-where-does-the-worlds-trash-go/258234/
9. www.economist.com/graphic-detail/2012/06/07/a-rubbish-map
10. www.climatedepot.com/2016/01/12/satellites-no-global-warming-at-all-for-18-years-8-months
11. www.principia-scientific.org/we-all-exhale-co2-at-40000-parts-per-million/
12. www.news.nationalgeographic.com/2017/07/plastic-produced-recycling-waste-ocean-trash-debris-environment/
13. www.google.com/search?q=glass%2C+metal%2C+and+plastic+recycling+costs+New+York+City+%24240+per+ton%2C&oq=glass%2C+metal%2C+and+plastic+recycling+costs+New+York+City+%24240+per+ton%2C&aqs=chrome..69i57.1721j0j9&sourceid=chrome&ie=UTF-8
14. www.scienceline.org/2008/05/ask-intagliata-plastic/
15. www.scienceline.org/2008/05/ask-intagliata-plastic/
16. www.earth-policy.org/mobile/books/pb4/PB4ch4_ss6?phpMyAdmin=1d6bec1fea35111307d869d19bcd2ce7
17. www.epa.gov/newsreleases/epa-proposes-affordable-clean-energy-ace-rule

Chapter 8

1. www.rubiconglobal.com/blog-brief-history-sustainability-movement/
2. www.macalester.edu/sustainability/zero-waste/
3. www.pdfs.semanticscholar.org/1c34/baac83524c4d4f65318674983c3dd0c027ca.pdf

Chapter 9

1. www.climate.nasa.gov/400ppmquotes/
2. www.democracynow.org/2013/5/13/climate_tipping_point_concentration_of_carbon
3. www.nature.com/news/2009/090109/full/news.2009.13.html
4. www.news.discovery.com/earth/plants/largest-iron-fertilization-test-blooms-criticism-120719.htm
5. www.mobile.nytimes.com/2013/01/06/business/pilot-plant-in-the-works-for-carbon-dioxide-cleansing.html?_r=0
6. www.newscientist.com/gallery/geoengineering
7. www.climatescience.org.au/content/378-e8-climate-stabilization-climate-change-commitment-and-irreversibility
8. www.bbc.com/news/science-environment-24033772

Chapter 10

1. www.repository.upenn.edu/cgi/viewcontent.cgi?article=1017&context=think_tanks
2. www.onlinelibrary.wiley.com/doi/full/10.1111/1758-5899.12295
3. https://www.climatechangenews.com/2016/11/18/cop22-headlines-what-did-marrakech-climate-summit-deliver/
4. www.wattsupwiththat.com/2011/05/29/its-all-over-kyoto-protocol-loses-four-big-nations/
5. www.bseec.org/renewable_energy_depends_on_fossil_fuels
6. www.forbes.com/sites/jamesconca/2017/05/30/why-do-federal-subsidies-make-renewable-energy-so-costly/#28ae5b06128c
7. www.youtube.com/watch?v=kHZKo13HV2A&t=642s

Chapter 11

1. www.climatecommunication.yale.edu/publications/politics-global-warming-march-2018/2/
2. www.pewinternet.org/2016/10/04/the-politics-of-climate/ps_2016-10-04_politics-of-climate_0-01/

3. www.forbes.com/sites/jamestaylor/2013/05/30/global-warming-alarmists-caught-doctoring-97-percent-consensus-claims/#2c2b17f8485d
4. www.forbes.com/sites/jamestaylor/2013/05/30/global-warming-alarmists-caught-doctoring-97-percent-consensus-claims/#2c2b17f8485d
5. www.wunderground.com/resources/climate/928.asp
6. www.ajc.com/news/science/more-than-half-americans-don-think-climate-change-will-affect-them/zQ5TseDhuJiXVCzHwoeEdM/
7. www.investors.com/politics/editorials/climate-skeptic-lennart-bengtsson-paper-suppressed/
8. www.investors.com/politics/editorials/climate-skeptic-lennart-bengtsson-paper-suppressed/
9. www.investors.com/politics/editorials/climate-skeptic-lennart-bengtsson-paper-suppressed/
10. www.telegraph.co.uk/news/earth/environment/climatechange/8786565/War-of-words-over-global-warming-as-Nobel-laureate-resigns-in-protest.html
11. www.telegraph.co.uk/news/earth/environment/climatechange/8786565/War-of-words-over-global-warming-as-Nobel-laureate-resigns-in-protest.html
12. www.sciencebits.com/IceCoreTruth
13. www.iheart.com/content/2017-08-03-youll-never-guess-how-many-homes-yearly-energy-output-al-gores-heated-pool-would-cover/
14. www.theclimategatebook.com/why-al-gore-refuses-to-debate-anyone/

Chapter 12

1. www.na.unep.net/geas/getuneppagewitharticleidscript.php?article_id=71
2. www.worldpopulationreview.com/world-cities/mumbai-population/
3. www.climatedepot.com/2013/01/10/paging-ehrlich-slate-mag-about-that-overpopulation-problem-research-suggests-we-may-actually-face-a-declining-world-population-in-the-coming-years/

4. www.newatlas.com/go/7334/
5. www.google.com/search?q=this+demographic+milestone+in+May+2007+and+became+more+urban+than+rural.&oq=this+demographic+milestone+in+May+2007+and+became+more+urban+than+rural.&aqs=chrome..69i57j69i64.1351j0j4&sourceid=chrome&ie=UTF-8
6. www.hrsa.gov/rural-health/about-us/definition/index.html
7. www.peopleandtheplanet.com/index.html@lid=27115§ion=40&topic=26.html
8. www.ourworldindata.org/extreme-poverty
9. www.ourworldindata.org/extreme-poverty
10. www.sciencedaily.com/releases/2007/05/070525000642.htm
11. www.worldhunger.org/world-hunger-and-poverty-facts-and-statistics/
12. www.britannica.com/biography/Thomas-Malthus
13. www.bbc.com/news/health-46118103
14. www.abcnews.go.com/Technology/story?id=98371&page=1

Conclusion

1. www.windows2universe.org/earth/Water/ocean_heat_storage_transfer.html
2. www.wattsupwiththat.com/2014/07/29/epa-document-supports-3-of-atmospheric-carbon-dioxide-is-attributable-to-human-sources/
3. www.pods.dasnr.okstate.edu/docushare/dsweb/Get/Document-10655/HLA-6723web.pdf
4. www.desmogblog.com/william-happer
5. www.explorable.com/falsifiability
6. www.wattsupwiththat.com/2014/01/01/ipcc-silently-slashes-its-global-warming-predictions-in-the-ar5-final-draft/
7. www.unglobalcompact.org/take-action/action/cop21-business-action
8. www.realclearpolitics.com/articles/2018/12/07/climate_change_alarmism_is_the_worlds_leading_cause_of_hot_gas_138852.html

9. www.prospect.org/article/holdrens-controversial-population-control-past
10. www.thespiritnewspaper.com/climate-change-alarmism-is-the-worlds-leading-cause-of-hot-gas-by-da-p11803-94.htm

REFERENCES

1. An Inconvenient Truth, Gore, A., 2006, Rodale Press
2. Climate Change, The Facts, Jennifer Marohasy, Editor, 2017 Institute of Public Affairs
3. Cool It, Lomborg, B, 2010, Vintage Books
4. Dark Winter, Casey, J.L., 2014, Humanix Books
5. Don't Sell Your Overcoat, Ambler, H, 2011, Lansing International Books
6. Earth In The Balance. Gore, A., 1992 Houghton Mifflin
7. Green Hell, Malloy, S, 2009, Blackstone Audio
8. Luke Warming, Michaels, P.J., Knappenberger,P.C., 2016, Cato Institute
9. Our Choice, Gore, A.,2009 Simon and Shuster
10. Storms of My Grandchildren, Hansen, J., 2009, Bloomsbury Publishing
11. The Population Bomb, Ehrlich. P, 1968, Sierra Club/ Ballentine Books

APPENDIX

Milestones in the Awareness of Global Warming

Courtesy of the American Institute of Physics

Here are gathered in chronological sequence many of the most important events in the history of climate change science.

1800-1870

Level of carbon dioxide gas (CO2) in the atmosphere, as later measured in ancient ice, is about 290 ppm (parts per million).

Mean global temperature (1850-1890) is roughly 13.7°C.

First Industrial Revolution. Coal, railroads, and land clearing speed up greenhouse gas emission, while better agriculture and sanitation speed up population growth.

1824

Fourier calculates that the Earth would be far colder if it lacked an atmosphere.

1859
Tyndall demonstrates that some gases block infrared radiation, and notes that changes in the concentration of the gases could bring climate change.

1879
International Meteorological Organization begins to compile and standardize global weather data, including temperature.

1896
Arrhenius publishes first calculation of global warming from human emissions of CO2.

1897
Chamberlin produces a model for global carbon exchange including feedbacks.

1870-1910
Second Industrial Revolution. Fertilizers and other chemicals, electricity, and public health further accelerate growth.

1914-1918
World War I; governments learn to mobilize and control industrial societies.

1920-1925
Opening of Texas and Persian Gulf oil fields inaugurates era of cheap energy.

1930s
Global warming trend since late nineteenth century reported.

Milankovitch proposes orbital changes as the cause of ice ages.

1938

Callendar argues that CO2 greenhouse global warming is underway, reviving interest in the question.

1939-1945

World War II. Military grand strategy is largely driven by a struggle to control oil fields.

1945

U.S. Office of Naval Research begins generous funding of many fields of science, some of which happen to be useful for understanding climate change.

1955

Phillips produces a convincing computer model of the global atmosphere.

1956

Ewing and Donn offer a feedback model for quick ice age onset.

Plass calculates that adding CO2 to the atmosphere will have a significant effect on the radiation balance.

1957

Launch of Soviet Sputnik satellite. Cold War concerns support 1957-58 International Geophysical Year, bringing new funding and coordination to climate studies.

Revelle finds that CO_2 produced by humans will not be readily absorbed by the oceans.

1958
Telescope studies show a greenhouse effect raises temperature of the atmosphere of Venus far above the boiling point of water.

1960
Mitchell reports downturn of global temperatures since the early 1940s.

Keeling accurately measures CO_2 in the Earth's atmosphere and detects an annual rise. The level is 315 ppm. Mean global temperature (five-year average) is 13.9°C.

1962
Cuban Missile Crisis, peak of the Cold War.

1963
Calculations suggest that feedback with water vapor could make the climate acutely sensitive to changes in CO_2 level.

1965
Boulder, Colorado meeting on causes of climate change: Lorenz and others point out the chaotic nature of climate system and the possibility of sudden shifts.

1966
Emiliani's analysis of deep-sea cores and Broecker's analysis of ancient corals show that the timing of ice ages was set by

small orbital shifts, suggesting that the climate system is sensitive to small changes.

1967

International Global Atmospheric Research Program established, mainly to gather data for better short-range weather prediction, but including climate.

Manabe and Wetherald make a convincing calculation that doubling CO2 would raise world temperatures a couple of degrees.

1968

Studies suggest a possibility of collapse of Antarctic ice sheets, which would raise sea levels catastrophically.

1969

Astronauts walk on the Moon, and people perceive the Earth as a fragile whole.

Budyko and Sellers present models of catastrophic ice-albedo feedbacks.

Nimbus III satellite begins to provide comprehensive global atmospheric temperature measurements.

1970

First Earth Day. Environmental movement attains strong influence, spreads concern about global degradation.

Creation of U.S. National Oceanic and Atmospheric Administration, the world's leading funder of climate research.

Aerosols from human activity are shown to be increasing swiftly. Bryson claims they counteract global warming and may bring serious cooling.

1971

SMIC conference of leading scientists reports a danger of rapid and serious global change caused by humans, calls for an organized research effort.

Mariner 9 spacecraft finds a great dust storm warming the atmosphere of Mars, plus indications of a radically different climate in the past.

1972

Ice cores and other evidence show big climate shifts in the past between relatively stable modes in the space of a thousand years or so, especially around 11,000 years ago.

Droughts in Africa, Ukraine, India cause world food crisis, spreading fears about climate change.

1973

Oil embargo and price rise bring first "energy crisis"

1974

Serious droughts since 1972 increase concern about climate, with cooling from aerosols suspected to be as likely as warming; scientists doubt all theories as journalists talk of a new ice age

1975
Warnings about environmental effects of airplanes lead to investigations of trace gases in the stratosphere and discovery of danger to ozone layer.

Manabe and collaborators produce complex but plausible computer models which show a temperature rise of a few degrees for doubled CO2.

1976
Studies show that CFCs (1975) and also methane and ozone (1976) can make a serious contribution to the greenhouse effect.

Deep-sea cores show a dominating influence from 100,000-year Milankovitch orbital changes, emphasizing the role of feedbacks.

Deforestation and other ecosystem changes are recognized as major factors in the future of the climate.

Eddy shows that there were prolonged periods without sunspots in past centuries, corresponding to cold periods.

1977
Scientific opinion tends to converge on global warming, not cooling, as the chief climate risk in the next century.

1978
Attempts to coordinate climate research in U.S. end with an inadequate National Climate Program Act, accompanied by rapid but temporary growth in funding.

1979

Second oil "energy crisis." Strengthened environmental movement encourages renewable energy sources, inhibits nuclear energy growth.

U.S. National Academy of Sciences report finds it highly credible that doubling CO_2 will bring 1.5-4.5°C global warming.

World Climate Research Programme launched to coordinate international research.

1981

Election of Reagan brings backlash against environmental movement to power. Political conservatism is linked to skepticism about global warming.

IBM Personal Computer introduced. Advanced economies are increasingly delinked from energy.

Hansen and others show that sulfate aerosols can significantly cool the climate, raising confidence in models that incorporate aerosols and show future greenhouse warming.

Some scientists predict greenhouse warming "signal" should become visible around the year 2000.

1982

Greenland ice cores reveal drastic temperature oscillations in the space of a century in the distant past.

Strong global warming since mid-1970s is reported, with 1981 the warmest year on record.

1983

Reports from U.S. National Academy of Sciences and Environmental Protection Agency spark conflict; greenhouse warming becomes a factor in mainstream politics.

Speculation over catastrophic climate change following a nuclear war or a dinosaur-killing asteroid strike promotes awareness of the atmosphere's fragility.

1985

Ramanathan and collaborators announce that global warming may come twice as fast as expected, from rise of methane and other trace greenhouse gases.

Villach Conference declares consensus among experts that some global warming seems inevitable, calls on governments to consider international agreements to restrict emissions.

Antarctic ice cores show that CO2 and temperature went up and down together through past ice ages, pointing to powerful feedbacks.

Broecker speculates that a reorganization of North Atlantic Ocean circulation can bring swift and radical climate change.

1986

Meltdown of reactor at Chernobyl (Soviet Union) cripples plans to replace fossil fuels with nuclear power.

1987

Montreal Protocol of the Vienna Convention imposes international restrictions on emission of ozone-destroying gases.

1988

News media coverage of global warming leaps upward following record heat and droughts plus statements by Hansen.

Toronto conference calls for strict, specific limits on greenhouse gas emissions; UK Prime Minister Thatcher is first major leader to call for action.

Ice-core and biology studies confirm living ecosystems give climate feedback by way of methane, which could accelerate global warming

Intergovernmental Panel on Climate Change (IPCC) is established.

1989

Fossil-fuel and other U.S. industries form Global Climate Coalition to tell politicians and the public that climate science is too uncertain to justify action.

1990

First IPCC report says world has been warming and future warming seems likely.

1991

Mt. Pinatubo explodes; Hansen predicts cooling pattern, verifying (by 1995) computer models of aerosol effects.

Global warming skeptics claim that twentieth century temperature changes followed from solar influences. (The solar-climate correlation would fail in the following decade.)

Studies from 55 million years ago show possibility of eruption of methane from the seabed with enormous self-sustained warming.

1992

Conference in Rio de Janeiro produces UN Framework Convention on Climate Change, but U.S. blocks calls for serious action.

Study of ancient climates reveals climate sensitivity to CO2 in same range as predicted independently by computer models.

1993

Greenland ice cores suggest that great climate changes (at least on a regional scale) can occur in the space of a single decade.

1995

Second IPCC report detects "signature" of human-caused greenhouse effect warming, declares that serious warming is likely in the coming century.

Reports of the breaking up of Antarctic ice shelves and other signs of actual current warming in polar regions begin affecting public opinion.

1997

Toyota introduces Prius in Japan, first mass-market electric hybrid car; swift progress in large wind turbines, solar electricity, and other energy alternatives.

International conference produces Kyoto Protocol, setting targets for industrialized nations to reduce greenhouse gas emissions if enough nations sign onto a treaty (rejected by U.S. Senate in advance).

1998

A "Super El Niño" makes this an exceptionally warm year, equaled in later years but not clearly exceeded until 2014. Borehole data confirm extraordinary warming trend.

Qualms about arbitrariness in computer models diminish as teams model ice-age climate and dispense with special adjustments to reproduce current climate.

1999

Criticism that satellite measurements show no warming are dismissed by National Academy Panel.

Ramanathan detects massive "brown cloud" of aerosols from South Asia.

2000

Global Climate Coalition dissolves as many corporations grapple with threat of warming, but oil lobby convinces U.S. administration to deny problem.

Variety of studies emphasize variability and importance of biological feedbacks in carbon cycle, liable to accelerate warming.

2001

Third IPCC report states baldly that global warming, unprecedented since the end of the last ice age, is "very likely," with highly damaging future impacts and possible severe surprises. Effective end of debate among all but a few scientists.

Bonn meeting, with participation of most countries but not U.S., develops mechanisms for working towards Kyoto targets.

National Academy panel sees a "paradigm shift" in scientific recognition of the risk of abrupt climate change (decade-scale).

Warming observed in ocean basins; match with computer models gives a clear signature of greenhouse effect warming.

2002

Studies find surprisingly strong "global dimming," due to pollution, has retarded arrival of greenhouse warming, but dimming is now decreasing.

2003

Numerous observations raise concern that collapse of ice sheets (West Antarctica, Greenland) can raise sea level faster than most had believed.

Deadly summer heat wave in Europe accelerates divergence between European and U.S. public opinion.

2004

First major books, movie, and art work featuring global warming appear.

2005

Kyoto treaty goes into effect, signed by major industrial nations except U.S. Work to retard emissions accelerates in Japan, Western Europe, U.S. regional governments and corporations.

Hurricane Katrina and other major tropical storms spur debate over impact of global warming on storm intensity.

2006

In longstanding "hockey stick" controversy, scientists conclude post-1980 global warming was unprecedented for centuries or more. The rise could not be attributed to changes in solar energy.

"An Inconvenient Truth" documentary persuades many but sharpens political polarization.

China overtakes the United States as the world's biggest emitter of CO2.

2007

Fourth IPCC report warns that serious effects of warming have become evident; cost of reducing emissions would be far less than the damage they will cause.

Greenland and Antarctic ice sheets and Arctic Ocean sea-ice cover found to be shrinking faster than expected.

2008

Climate scientists (although not the public) recognize that even if all greenhouse gas emissions could be halted immediately, global warming will continue for millennia.

2009

Many experts warn that global warming is arriving at a faster and more dangerous pace than anticipated just a few years earlier.

Excerpts from stolen e-mails of climate scientists fuel public skepticism.

Copenhagen conference fails to negotiate binding agreements: end of hopes of avoiding dangerous future climate change.

2011

Reaction to nuclear reactor disaster at Fukushima (Japan) ends hopes for a renaissance of nuclear power.

2012

Controversial "attribution" studies find recent disastrous heat waves, droughts, extremes of precipitation, and floods were made worse by global warming.

2013

An apparent pause or "hiatus" in global warming of the atmosphere since 1998 is explained; the world is still warming (as the next three record-breaking years would confirm).

2015

Researchers find collapse of West Antarctic ice sheet is irreversible, will bring meters of sea-level rise over future centuries.

Paris Agreement: nearly all nations pledge to set targets for their own greenhouse gas cuts and to report their progress.

Solar electricity and wind power become economically competitive with fossil fuels in some regions.

Mean global temperature is 14.8°C, the warmest in thousands of years. Level of CO2 in the atmosphere goes above 400 ppm, the highest in millions of years.

Printed in the United States
By Bookmasters